**COMPUTER
SOFTWARE APPLICATIONS
IN CHEMISTRY**

COMPUTER SOFTWARE APPLICATIONS IN CHEMISTRY

Second Edition

PETER C. JURS
Department of Chemistry
The Pennsylvania State University
University Park, Pennsylvania

A Wiley-Interscience Publication
JOHN WILEY & SONS, INC.
New York · Chichester · Brisbane · Toronto · Singapore

QD
39.3
E46
J873
1996

Library of Congress Cataloging in Publication Data:

Jurs, Peter C.
    Computer software applications in chemistry / Peter C. Jurs.—
2nd ed.
        p.   cm.
    Includes bibliographical references and index.
    ISBN 0-471-10587-2 (cloth : alk. paper)
    1. Chemistry—Data processing.   I. Title.
QD39.3.E46J873   1996
542'.85'53—dc20   95-41349
                                                        CIP

Printed in the United States of America

10 9 8 7 6 5 4 3 2 1

CONTENTS

PREFACE

Computers continue to be an integral part of everyday life, with an ever-increasing presence, and the conduct of science has been and is as heavily influenced as any area of human endeavor. The need for computer skills on the part of practicing chemists continues to grow. To meet this need, computational material is included in many chemistry textbooks and courses along with more traditional material. Books for new courses that go beyond such add-ons are also needed, and the present volume serves that purpose.

Over the past forty years, the two major subdivisions of computer applications in chemistry have been (1) hardware interfacing and laboratory usage, where the computer becomes part of the instrument, and (2) software usage involving numerical computations and abstract information processing tasks. More recently, the physical sciences have seen the development of powerful simulation methods. Such approaches are now so common and so useful to the advancement of science that many speak of the addition of simulation to the two traditional tools of experimentation and theory to produce a new era of three major contributors to scientific inquiry and advancement. This book focuses on software usage, both numerical and nonnumerical, and provides some introduction to the area of simulation in a few selected areas.

The aim of this second edition remains to provide an overview or survey of a number of important software applications in chemistry. The student who uses this book is expected to know FORTRAN already; this book does not teach the language, but uses it as a vehicle for teaching. The coverage of each topic starts at an introductory level and then advances to practical uses for the methods described. Advanced material is included in selected subject areas. In addition to the topics included in the first edition, a number of new, timely topics have been added to this second edition, including the following: regression analysis, principal-components analysis, Monte Carlo integration, graph theory and the development of topological indices, molecular dynamics, and graphical depiction of chemical data and of molecular structures. Each chapter provides a list of leading references to other textbooks, review articles, and the primary scientific literature so that the student can locate more advanced discussion on any topic. Many chapters contain fully documented FORTRAN programs with sample input and output streams that illustrate the concepts being discussed. The intent is to provide sufficient information for the student to implement and run these programs independently. The routines can be used as a starting point for customization or enhancement to solve real chemical problems. Often these illustrative programs perform functions that would be useful to a practicing chemist solving problems beyond those used merely for illustration.

This book is intended primarily to be used as a textbook in a formal course at the advanced undergraduate or beginning graduate level. It can also be employed for professional development. Implementing the programs should be an expected part of either approach to the book's use. FORTRAN is used in this second edition once again because it continues to be widely used for scientific computation, an enormous backlog of FORTRAN software exists, and because it is commonly taught and used by students in college and university courses. With all its faults, FORTRAN continues to be an important presence in the scientific programming arena.

As with many skills, the best way to learn how to write and modify software to achieve the results desired is to do it. Thus, this book should be a starting place for those who wish to develop skills working with software in science.

PETER C. JURS

1

INTRODUCTION

Over the past 50 years the invention, the evolution, and the maturing of computers of many types—personal computers, minicomputers, supercomputers, workstations, parallel computers—have ushered in a new age. We are in the *computer age* and the *information society.* Science, like so many other aspects of human affairs, has been transformed by the availability of this uniquely powerful new tool.

Several generations of computers have evolved over this 50-year time period (Clementi 1989). *First*-generation computers were characterized by vacuum tubes and static memory and were hand-operated (1946–1955). The early manually assembled computers such as ENIAC (electronic numerical integrator and calculator) and later commercial products such as the UNIVAC I and the IBM-650 are examples of first-generation machines. *Second*-generation machines incorporated transistors, magnetic core memory, magnetic tapes, and punched cards (1955–1965). The IBM-3090 is an example of a second-generation machine. *Third*-generation machines contained integrated circuits, magnetic disks, system software, virtual memory, and time-sharing (1965–1975). The IBM-360 family of machines exemplify this type of computer. *Fourth*-generation machines incorporated VLSI (very large-scale integrated circuit) logic, large memories, high speeds, and a wide spectrum of machine sizes and capabilities, from supercomputers to personal computers (1975–1990). *Fifth*-generation computers are usually characterized by the wide-scale introduction of artificial intelligence (1984–?). This effort, started with the 1984 Japanese MITI Fifth Generation Project, is now in progress, and the outcome is not yet clearly visible. *Sixth*-generation machines are progressing along a different path, with massively parallel computation, very large memories, very high-speed computations, highly connected systems, workstation "farms," increasingly powerful personal computers, and networking (1990–?). Areas of current emphasis expected to impact near-future computer technology (and perhaps lead to further generations of computers) include neural technology, optical disks, integration of artificial-intelligence (AI) methods with application software, new materials for construction of logic circuits, increasingly larger-scale integration of circuit components, and advances in graphics and visualization methods.

1.1 CHARACTERISTICS OF DIGITAL COMPUTERS

Modern digital computers are uniquely capable devices. They can manipulate and transform information to carry out functions that were previously performed only by the human brain. The power and breadth of application of digital computers is a result of the fact that computers simultaneously possess a number of mutually supportive properties. The following paragraphs discuss these properties individually, but it is the combination at the same time in the same machine that leads to the uniqueness of the modern digital computer.

High-Speed Operation

High-speed operation is the property most often associated with computers. Although speed is an important factor in computer operation, speed alone is useless without associated properties as well. Nonetheless, very large computations can be performed routinely with fast machines. It is routine for computers to perform millions of operations per second. It is very difficult to comprehend the quantity one million, so we must resort to analogies and comparisons, as follow. There are approximately as many microseconds in one hour as there have been full seconds in the 125 years that have passed since the end of the American Civil War in 1865.

One million floating-point operations per second (1 Mflop/s) is no longer attainable by only fast computers, but is routinely available. Many workstations of the mid-1990s achieve speeds in the 5–30-Mflop/s range. Supercomputers such as those available from CRAY, NEC, and Fujitsu now have speeds in the hundreds of Mflops/s. Routine access to computers with speeds in the tens to hundreds of Mflops/s makes possible a myriad of computations, simulations, and other operations that were unrealistic only a few years ago. Accordingly, whole new areas of computational chemistry have blossomed, and this trend has no end in sight as computer speeds continue to increase.

Low Error Rate

The *low-error-rate* property refers to the fact that computers rarely (in fact, almost never) introduce errors into their operations through malfunctions. This is due primarily to high-reliability components and partly to system design and software design that continually checks for such errors and announces them when they do occur. This type of error must be distinguished from errors that result from poor user software design, as will be discussed in a later section.

Varied Representation of Information

Many people consider computers as merely "fast adding machines." While computers can do arithmetic computations extremely quickly, they are in fact much more than numerical calculators. Any information whatsoever can be rendered into a form suitable for storage, manipulation, retrieval, and display by digital computers. Thus, they are universal information processors.

There are usually at least two different types of numerical information in digital computers: integer (fixed-point) and real (floating-point). In the integer storage mode a fixed number of bits (often a multiple of 8 since computers are often designed with 8-bit bytes as the building blocks) are allocated for storage of a value. One bit is set aside for the sign, leaving the remaining bits for representing the absolute value. In a typical computer, 4 bytes, or 32 bits, constitute the default number assigned to integer storage, so 31 bits would be allocated to storage of the value. Thus, numbers in the range -2^{31} to $2^{31}-1$ can be stored. The precision of storage of a number in the integer mode depends on the absolute value of the number.

The real-storage mode achieves two important goals—scaling and constant precision—and uses a logarithmic notation much like scientific notation. In a typical computer, 4 bytes, or 32 bits, are assigned to a single-precision real value. One bit is allocated for the sign of the value, 7 bits for the exponent, and 24 bits for the fraction. In this representation, numbers between $\sim 2.4 \times 10^{-78}$ and 7.2×10^{75} can be stored, each with a constant precision of 1 part in 2^{24}. The scientific notation actually used in many computer systems is based on the octal or hexadecimal number systems. This is because numbers expressed in octal or hexadecimal are easily convertible to binary; Table 1.1 shows some numbers expressed in their binary, octal, decimal, and hexadecimal forms. A more detailed discussion of the floating-point number system is presented in a subsequent section.

General alphanumeric information can be stored as codes. Two of the best known such codes are EBCDIC (Extended Binary Coded Decimal Interchange Code) and ASCII (American Standard Code for Information Interchange). Each of these uses 2-byte, or 16-bit, patterns to represent each letter, number, and special character that must be coded. The actual code patterns can be found in many standard references. Thus, any alphanumeric information can be stored and therefore manipulated as well.

Stored Program Capability

Stored program capability frees the digital computer from the slow world of manual data processing as it executes its instructions. Once the sequence of

TABLE 1.1   Number System Bases and Conversions

Binary	Octal	Decimal	Hexadecimal
0	0	0	0
1	1	1	1
10	2	2	
11	3	3	3
100	4	4	4
101	5	5	5
110	6	6	6
111	7	7	7
1000	10	8	8
1001	11	9	9
1010	12	10	A
1011	13	11	B
1100	14	12	C
1101	15	13	D
1110	16	14	E
1111	17	15	F
10000	20	16	10
10001	21	17	11

instructions to be executed is stored internally, the instructions can be performed at electronic speed. This is the difference between a hand-held calculator performing calculations as the user punches buttons one at a time and a computer executing an entire program. The idea of storing the program was an insight of John von Neumann, one of the pioneers of computation, in the 1940s. He correctly saw that once the instructions were stored they could be operated on in the same manner as could data. This insight led von Neumann directly to the concept of conditional transfer, the next crucial property of digital computers.

Conditional Transfer

Conditional transfer means that an executing program can test the values of intermediate results, and it can then take one of several actions depending on the results of the test. This allows computer programs to break out of lock-step, routine execution of fully determined sequences of statements. It is conditional transfer that allows computers to behave flexibly. The recognition of the importance of conditional transfer was one of the breakthroughs of early computational theory that led directly to the modern digital computer and its enormous utility.

Digital Operation

The development of modern digital computers has been coupled with an emphasis on discrete-function mathematics. It has forced a rethinking of much of mathematics, which was previously concerned much more with continuous functions than with discrete functions. The standard paradigm of mathematics has been the application of calculus and its continuous functions to real-world problems. Often, this requires making simplifying approximations. It also often involves converting the calculus formation back into a discrete form for numerical solution. In contrast, digital computers operate inherently in the digital domain, and many problems to be tackled can be formulated directly in a digital framework. This has led to a rebirth of discrete-function mathematics (Ralston 1986).

Electronics and Technology

The basic data-processing concepts incorporated into computers existed well before they could be rendered in a practical device. Then, fueled by the necessities of World War II and the availability of vacuum tubes, electronics made computer implementation possible. The developments of computer technology can be divided into generations according to the dominant technologies. Starting with vacuum tubes, then transistors, then integrated circuits, and large-scale integrated circuits (VLSI), electronics technology has advanced the capability of computers. The basic deisgn of most computers consists of one central processing unit (CPU) connected to a main memory. This makes such computers inherently serial (as opposed to parallel) devices, although the rate of operations can be very fast. Such serial machines have been termed *von Neumann machines*. Fifth-generation computers are designed them without this inherent serial nature and incorporate artificial intelligence in the systems. Thus, while the earlier generations of computers relied on electronic advances, fifth- and future-generations computers will rely more heavily on new ideas for the architecture of the systems. In addition to electronics, other scientific and technological areas have contributed to advances in computer design, such as crystallography, optics, plasma physics, polymer chemistry, semiconductor technology, and surface physics and chemistry. In fact, the relationship between computer capability and architecture and IC technology continues to evolve and advance along many fronts, with the result that computers available on the market are becoming increasingly cost-effective.

1.2   SCIENTIFIC COMPUTER USES

Digital computers are being used in nearly all phases of modern science, and chemistry is certainly no exception. The infusion of computer methodology into science has occurred by degrees since the 1950s. Computers are now used in many phases of scientific experimentation and inquiry, and are so commonplace in many phases of science that their importance may seem blurred. Some of the most prominent applications of digital computers in chemistry, described in the following paragraphs, reflect the breadth of usage and importance of these new scientific instruments.

Numerical Calculations

The first and most obvious use of computers was in numerical calculations, and this application continues to be extremely important. The availability of powerful computers capable of performing extremely large computations has altered the nature of a number of scientific fields, including weather forecasting, nuclear reactor design and control, economic modeling, control of weapons systems, manned spacecraft design and use, environmental monitoring, cryptography, and telephone switching networks. Many calculations performed routinely now were inconceivable prior to the availability of large computers. For instance, compare the rate of computation by a person with that by a computer. Consider a hypothetical computer that can perform one million operations per second (1 Mflop/s). Multiply this by 3600 s/h (seconds per hour) to get 3.6×10^9 operations/per hour. For a person, assume an arithmetic computation rate of one operation/per second. Multiplying this by 3600 s/h, 8 h/day, 5 days/week, 50 weeks/year, and by 70 years to get 4×10^8 operations per lifetime. Thus, there is a very rough equivalence between one hour of machine time at 1 Mflop/s and one person's lifetime.

In chemistry, there are many areas in which the capability to perform large numbers of computations has become crucial. Examples of such fields include x-ray crystallography, quantum mechanics, neutron scattering, liquid simulations, reaction dynamics, and many others.

Time-Limited Problems

The dividing line between numerical calculations and time-limited problems is indistinct. However, some modern chemical experiments must have fast computations coupled to them in order to allow them to be attempted at all. The nature of many problems requires this.

For example, to perform Fourier transform spectroscopy [such as Fourier transform infrared (IR) spectroscopy or Fourier transform nuclear magnetic resonance (NMR) spectroscopy] requires the rapid transformation of large data arrays from the time domain to the frequency domain. The techniques of Fourier transform spectroscopy were all known well in advance of the availability of fast digital computers, but the techniques were not generally feasible. These techniques were implemented on a wide-scale basis only after development of inexpensive, fast-computing hardware and the discovery of the fast Fourier transform, a software breakthrough.

Modeling, Simulation, and Optimization

Modeling and simulation of chemical experiments, instruments, and systems is now routinely done to understand the interactions between the adjustable parameters and to search for the best values for them. Computers can be used in simulations to design chemical experiments or instruments and then to optimize them with respect to the adjustable parameters. This can be markedly more economical than building prototype experiments or instruments to be used for optimization.

Information Storage and Retrieval

Chemistry is often characterized as the scientific field with the best information organization. Within the domain of bibliographic information, this is largely because of the American Chemical Society's Chemical Abstracts Service (CAS), which has systematically abstracted the worldwide chemical literature. In the endeavor, CAS has come to employ computerization to keep up with the enormous demand for such services. The type of information stored by CAS is both structural and bibliographic. Bibliographic information consists of titles, authors, journal citations, abstracts of papers and patents, and so on. The structural information stored lists the molecular structures of the compounds reported in the abstracted literature. The computer software used to print *Chemical Abstracts* and many of the ACS journals must handle the structural representations used internally by the computers. Other large banks of computer-compatible structural representations and data exist in the public sector and within industrial companies.

The second major type of information repository is spectral databases, in which the data stored, manipulated, retrieved, and displayed are spectra of compounds such as mass spectra, IR spectra, or proton or ^{13}C NMR spectra. A number of commercial services offer software that allows searching of such databases by customers who are working on structure elucidation problems.

Experiment Management and Control

Most modern chemical instrumentation contains a computer as an integral part. This feature may be advertised as an advantage, or the computer may be buried within the instrument architecture so that it is hidden from the user. It is obvious that modern instrumental chemistry has become utterly dependent on computerization. Early uses of computers in instrumentation included datalogging capabilities alone. More recently, the microcomputers or minicomputers embedded within instruments have been endowed with a modicum of intelligence that allows some error checking, routine recalibration, adjusting for blank spectra or dark currents, and other such necessary steps in instrumental usage. This has led to the term *computer-aided chemistry* as applied to the laboratory.

Intelligent Problem Solving

Computers are certainly capable of performing prodigious amounts of computation, and this characteristic leads to important consequences in the way in which science is now conducted. However, computers are much more than fast adding machines; they are universal information processors and thus can be used for many intrinsically nonnumerical tasks. Computers can certainly be used to tackle tasks that are normally regarded as requiring human intelligence. Perhaps the primary chemical example is that of elucidation of chemical structure from spectroscopic and other data. This can be done partly by library searching of databases, but it also requires the use of high-level inference and true intelligence. There have been several research projects tackling this area of computer usage in chemistry, and this area is certain to grow dramatically in the decades ahead.

Graphical Display of Data and Molecular Structures

When computers are provided with graphical display devices, they can provide the user with new ways of seeing the results of computations or experiments. The display of complex data in graphical form is not new, but the ability to manipulate, scale, and rotate the display, and use hidden line plots and so on is new. These abilities are important in allowing the experimenter to deal interactively with the results, and they make the complete cycle of theory, experiment, and theory go much faster. The ability to display molecular structures on computer-controlled displays is really new. For the first time, chemists can view complex molecular structures while they are rotated, scaled, put in contact with other structures, and so on. There are many things that physical models of structures cannot be used for that are routine in computer displays. For example, with a computer display of an enzyme and its substrate, the user can investigate enzyme–substrate as a function of the relative locations of the two molecules. If unimportant sidechains in either of the molecules obscure the view, these portions of the structures could be suppressed. The chemists can tour the inside of large marcromolecules such as DNA (deoxyribonucleic acid) in a way never before possible.

1.3 ALGORITHM DESIGN

The most important aspect of applying computers to the solution of problems is that this task forces the scientist to completely define, fully analyze, and thoroughly understand the problem. The first step in attacking a problem is *problem analysis*. This involves defining what is to be done and how. It may involve breaking the original problem down into smaller problems. It will involve listing all the available inputs and all the desired outputs.

An *algorithm* is a procedure consisting of a well-defined, finite set of unambiguous rules giving a sequence of operations for solving a specific type of problem. Development of algorithms and studies of the relative merits of alternative algorithms for the solution of a given problem is one of the major areas of computer applications in science. Unlike the field of mathematics, existence theorems are not sufficient in computing. There must exist a work-

tines is the International Mathematical Statistical Library (IMSL 1979). Other similar packages also exist.

Recommendation 2. Thoroughly test programs over the domain of possible values for input data, especially on the limits of each domain. Modules within large programs should be tested individually.

Recommendation 3. Use double-precision arithmetic unless a program's numerical computations can be proved stable in single precision.

Recommendation 4. For a given level of computer precision, do not input or output more decimal digits than are meaningful.

Recommendation 5. Use relative comparisons in place of absolute comparisons. Never use the FORTRAN relational operators .EQ. or .NE. for comparing floating-point quantities. Tolerant comparisons can be constructed as follows:

```
TEQ(U,V)=DABS(X-Y).LE.DMAX1(DABS(X),DABS(Y))*EPS3
```

where EPS3 is the number 3*EPS; it allows the last two mantissa bits of any floating-point numbers x and y to be ignored during comparisons. This is because 3 base 10 = 11 base 2. Function TEQ means, "If the absolute difference between double-precision arguments corresponding to parameters X and Y, when scaled relative to the unit interval, is less than EPS adjusted to ignore the two low-order mantissa bits, then the numbers are computationally equal." Thus, TEQ(X,Y) tests for *computational equivalence,* which is presumably the intention of the program in any case. TEQ is a logical function, and the variables X, Y, and EPS3 should all be declared as double-precision variables in the program. Tolerant comparisons can be similarly constructed for .NE., .GT., and the other relational operators.

Recommendation 6. Develop numeric algorithms that are computationally robust. That is, write programs that either avoid significance loss or else monitor it. Knoble (1979) gives an example of a FORTRAN program that incorporates robust addition and subtraction routines.

REFERENCES

Aird, T. J., "The IMSL Fortran Converter. An Approach to Solving Portability Problems," in *Lecture Notes in Computer Science,* Springer-Verlag, New York, 1978.

Bevington, P. R., *Data Reduction and Error Analysis for the Physical Sciences,* McGraw-Hill, New York, 1969.

IMSL, *International Mathematical Statistical Library Reference Manual,* IMSL, Houston, TX, 1979.

Knoble, H. D., *A Practical Look at Computer Arithmetic,* Computation Center, The Pennsylvania State University, Aug. 1979.

Knuth, D. E., *The Art of Computer Programming,* Vol. 2, Addison-Wesley, Reading, MA, 1969, Sections 4.2 and 4.3.

Norris, A. C., *Computational Chemistry, an Introduction to Numerical Methods,* Wiley, New York, 1981.

Solberg, H. E., "Inaccuracies in Computer Calculation of Standard Deviation," *Anal. Chem.,* **55,** 1611 (1983).

Wanek, P. M., et al., "Inaccuracies in the Calculation of Standard Deviation with Electronic Calculators, *Anal. Chem.,* **54,** 1877 (1982).

$$e_{is} = \frac{e_i}{s} \qquad (3.33)$$

The standardized residuals have zero mean and unit standard deviations.

The residuals should be distributed according to the normal distribution. That is, the standardized residuals should have random values distributed about zero with 95% of the values falling between -2 and $+2$. If the model that has been generated is wrong, then a plot of the residuals will often display the fact plainly. Thus, making plots of residuals and analyzing them is a powerful tool for determining whether models are good representations of your data.

Such plots can be constructed in many ways, but several popular variations are as follows: (1) plot e_{is} versus the model prediction y_i; (2) plot e_{is} versus the independent variable x_i; and (3) plot e_{is} versus another possibly relevant experimental variable such as time. When plots such as these are constructed, they should show no patterns among the residuals. They should appear scattered and random if the model is a good representation of the data. Examination of residuals is one of the most generally useful tools of analysis in regression studies.

Example of a Linear Fit   A set of 10 points is to be fit with a linear equation. The points are (0.4, 1.217), (0.8, 1.821), (1.25, 2.715), (1.6, 2.835), (2.0, 3.544), (2.5, 4.778), (3.1, 5.721), (3.5, 6.179), (4.0, 7.266), and (4.4, 7.451). The summations necessary to calculate the intercept and slope are

$$\Sigma x_i = 23.55 \qquad \Sigma y_i = 43.53$$

$$\Sigma x_i^2 = 72.39 \qquad \Sigma x_i y_i = 130.12$$

Then the intercept and slope and their deviations are

$$a_0 = 0.51 \pm 0.13 \qquad a_1 = 1.63 \pm 0.05$$

The value of the SSE is 0.33, and the value of the linear correlation coefficient is $R = 0.99$. The fit equation is

$$y = 0.51 + 1.63x$$

A plot of the 10 points and the best-fit line is shown in Figure 3.3.

Example of Difficulties with Linear Fits   The statistical quantities SSE and R can be misleading as to the quality of a linear fit to a set of data. The following

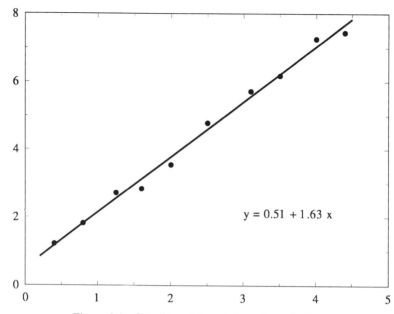

Figure 3.3   Set of ten data points and best-fit line.

set of data, taken from a paper by Anscombe (1973), illustrates this point from several viewpoints.

The set of data to be used for this exercise consists of four pairs of x–y data as follows:

x_1	y_1	x_2	y_2	x_3	y_3	x_4	y_4
10	8.04	10	9.14	10	7.46	8	6.58
8	6.95	8	8.14	8	6.77	8	5.76
13	7.58	13	8.74	13	12.74	8	7.71
9	8.81	9	8.77	9	7.11	8	8.84
11	8.33	11	9.26	11	7.81	8	8.47
14	9.96	14	8.10	14	8.84	8	7.04
6	7.24	6	6.13	6	6.08	8	5.25
4	4.26	4	3.10	4	5.39	19	12.50
12	10.84	12	9.13	12	8.15	8	5.56
7	4.82	7	7.26	7	6.42	8	7.91
5	5.68	5	4.74	5	5.73	8	6.89

A least-squares fit of a straight line through any of these four sets of data points leads to exactly the same equation, $y = 3 + \frac{1}{2}x$, with exactly the same goodness-of-fit parameters; that is, $R^2 = 0.67$ and $s = 1.24$ for each set of points. It isn't until you look at the scatterplots of the data that the absurdity of the situation is clearly revealed.

Figure 3.4a shows a scatterplot of y_1 versus x_1. These data do, indeed, fit a straight-line model, although they are quite scattered about the fit line. Figure 3.4b shows a scatterplot of y_2 versus x_2. These data do not agree with a straight-line model at all well. The data have obvious curvature, and they should be fit with a quadratic equation. In fact, when this is done it turns out that these 10 data points fit exactly to a quadratic equation. Figure 3.4c shows a scatterplot of y_3 versus x_3. These data do fit a straight-line model, but there is one extreme outlier point that substantially affects the coefficients of the fit equation. Figure 3.4d shows a scatterplot of y_4 versus x_4. Nine of these data points are clustered at the value of $x = 8$, and only one point has a different x value. It is nonsensical to fit these data with any kind of equation using least squares.

Many lessons are apparent from these sets of data and the linear fits shown in the figures. The most obvious point is that one must look at graphs of the data with the fit equation superimposed in order to see many types of potential errors. You cannot rely completely on goodness of fit parameters, because they can be in error.

3.2 GENERAL POLYNOMIAL EQUATION FITTING

To this point, we have dealt with simplest linear equation as the function to be fit. The same approach can be used to fit polynomials of higher orders to data. The general polynomial of order m is

$$f(x_i) = \sum_{j=1}^{m} a_j x_i^j \qquad (3.34)$$

Substituting this function form into Equation (3.16) yields

$$Q = \sum_i w_i \left[y_i - \sum_j a_j x_i^j \right]^2 \qquad (3.35)$$

The objective is to find the values for the a_j that minimize Q. To minimize, we differentiate with respect to each adjustable parameter and set it equal to zero to get a set of normal equations,

$$\frac{\partial Q}{\partial a_k} = 0 = \sum_i w_i \left[y_i - \sum_j a_j x_i^j \right] x_i^k \qquad k = 0,1,2, \ldots , m \qquad (3.36)$$

We have a set of $m + 1$ equations in $m + 1$ unknowns, so they can be solved. The equations can be most easily expressed in matrix form, so this way of defining the problem will now be developed.

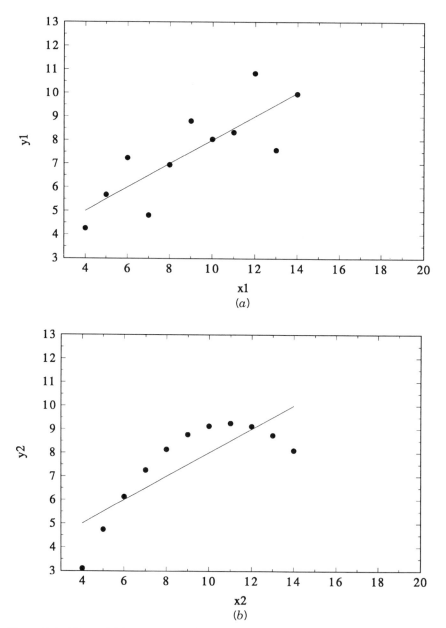

Figure 3.4   Plots of (a) y_1 versus x_1, (b) y_2 versus x_2, (c) y_3 versus x_3, (d) y_4 versus x_4 with linear least-squares fit shown.

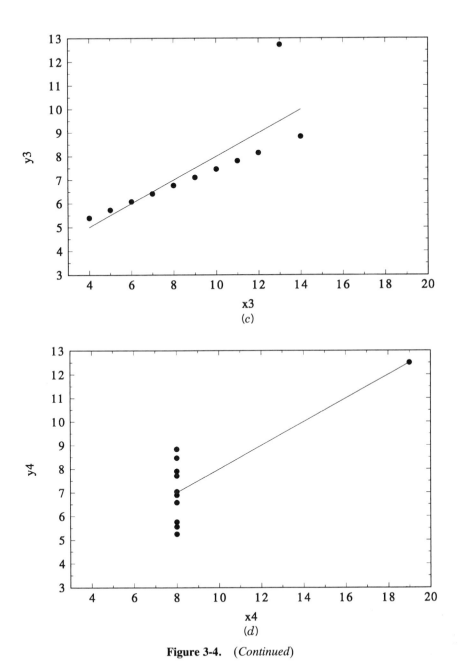

Figure 3-4.   (*Continued*)

As a specific case, we take the polynomial equation of degree 2, the quadratic:

$$f(x_i) = b_0 + b_1 x_i + b_2 x_i^2 \tag{3.37}$$

In terms of this function, the normal equations become

$$\sum_i w_i(y_i - b_0 - b_1 x_i - b_2 x_i^2) x_i^k = 0 \quad k=0,1,2 \tag{3.38}$$

Expanded, these become

$$
\begin{aligned}
b_0 \, \Sigma \, w_i \;\;\;\; + b_1 \, \Sigma \, w_i \, x_i + b_2 \, \Sigma \, w_i \, x_i^2 &= \Sigma \, w_i \, y_i \\
b_0 \, \Sigma \, w_i \, x_i + b_1 \, \Sigma \, w_i \, x_i^2 + b_2 \, \Sigma \, w_i \, x_i^3 &= \Sigma \, w_i \, x_i \, y_i \\
b_0 \, \Sigma \, w_i \, x_i^2 + b_1 \, \Sigma \, w_i \, x_i^3 + b_2 \, \Sigma \, w_i \, x_i^4 &= \Sigma \, w_i \, x_{ii}^2
\end{aligned}
\tag{3.39}
$$

This set of equations can be written in matrix form as

$$\mathbf{A}\,\mathbf{B} = \mathbf{C} \tag{3.40}$$

$$\mathbf{A} = \begin{pmatrix} \Sigma \, w_i & \Sigma \, w_i x_i & \Sigma \, w_i x_i^2 \\ \Sigma \, w_i x_i & \Sigma \, w_i x_i^2 & \Sigma \, w_i x_i^3 \\ \Sigma \, w_i x_i^2 & \Sigma \, w_i x_i^3 & \Sigma \, w_i x_i^4 \end{pmatrix}$$

$$\mathbf{B} = \begin{pmatrix} b_0 \\ b_1 \\ b_2 \end{pmatrix}$$

$$\mathbf{C} = \begin{pmatrix} \Sigma \, w_i y_i \\ \Sigma \, w_i x_i y_i \\ \Sigma \, w_i x_i^2 y_i \end{pmatrix}$$

To solve this set of linear equations, it is necessary to take the inverse of the coefficient matrix $\mathbf{A}$. Give this new matrix the name $\mathbf{D}$ and call its elements d_{ij}. Then, by definition,

$$\mathbf{B} = \mathbf{A}^{-1}\,\mathbf{C} = \mathbf{D}\,\mathbf{C} \tag{3.41}$$

The values in the $\mathbf{B}$ matrix, namely b_0, b_1, and b_2, are the least-squares best values for the three coefficients of the quadratic fit sought.

The standard deviation of the fit s is calculated from the equation

$$s^2 = \frac{Q}{n - p - 1} \tag{3.42}$$

where Q comes from Equation (3.35), n is the number of data points being fit, and p is the degree of the polynomial. Note that Equation (3.42) reduces to Equation (3.21) for the linear equation where $p = 1$. Now the standard deviations for each of the parameters can be computed in terms of s^2 as

$$s_{b_0}^2 = d_{11}s^2 \qquad s_{b_1}^2 = d_{22}s^2 \qquad s_{b_2}^2 = d_{33}s^2 \tag{3.43}$$

where the d_{kk} values are the elements from the main diagonal of the inverse matrix **D**.

These equations that have been derived for the quadratic equation are applicable for higher-order polynomials as well. In general, one must invert a $(p + 1) \times (p + 1)$ matrix, where p is the degree of the polynomial. The equations for the standard deviations of the fit parameters also generalize to higher order.

A word of caution is in order to finish this discussion of polynomial curve fitting. Many computer libraries and many software packages available to users contain routines that will take a set of data and fit polynomial equations to it in an iterative way. That is, the degree of the polynomial is increased by 1 after each fit, and the fitting is done again. Comparative statistics for the degree of goodness of the fit as a function of the polynomial degree are output. The user must carefully avoid using such routines incorrectly because it is common that a higher-order polynomial will provide a better quality fit than will a lower-order one, but without additional true meaning. Just because a higher-order polynomial can fluctuate more, it can sometimes fit noisy data better than a lower order equation. A rule of thumb that is widely practiced is to keep the degree of the polynomial being fit to less than one-third or even one-fifth of the number of data points being fit. Thus, a data set of 20 points should probably not be fit with more than about a cubic equation to be on the safe side.

3.3 LINEARIZING TRANSFORMATIONS

The equations to be fit to data are not always polynomials, let alone straight lines. Many other functional forms arise in scientific investigations, and they must be dealt with. One approach for dealing with some functional forms is to use a transformation to make the curve-fitting problem into a linear one. An example function that can be fit using this strategy is the exponential function

$$f(x_i) = y_i = \alpha \exp(\beta x_i) \tag{3.44}$$

If $\beta > 0$, this is a rising exponential curve (like a plot of the earth's population vs. time), and if $\beta < 0$, the curve is a decaying exponential (like a plot of radioactive count rate vs. time for a decaying radioactive species). Taking natural logarithms of both sides of Equation (3.44) yields

$$y_i' = \ln y_i = \ln \alpha + \beta x_i \tag{3.45}$$

which is a linear equation after the substitutions $y_i = \ln y_i$, $a_0 = \ln \alpha$, and $a_1 = \beta$. Now

$$y_i' = a_0 + a_1 x_i \tag{3.46}$$

To maintain the proper relationship between the weights and the points, we must also transform the weights. The general equation for propagation of error is

$$\sigma_y^2 = \sum_i \sigma_{x_i}^2 \left[\frac{\partial y}{\partial x_i} \right]^2 \tag{3.47}$$

for any function $y = f(x_i)$. Now, the weights can normally be considered inversely related to the standard deviations of the points

$$w_i = \frac{1}{\sigma_{y_i}^2} \tag{3.48}$$

Then the weights are transformed using the general error propagation equation above to get

$$w_i' = w_i \left[\frac{\partial y_i'}{\partial y_i} \right]^{-2} \tag{3.49}$$

and

$$\frac{\partial y_i'}{\partial y_i} = \frac{1}{y_i} \tag{3.50}$$

so

$$w_i' = w_i y_i^2 \tag{3.51}$$

The transformations from the original data points (x_i, y_i) with associated weights w_i to the new linear problem are

$$y_i' = ln\ y_i \quad \text{and} \quad w_i' = w_i y_i^2 \tag{3.52}$$

Using the transformed points and weights, a linear fitting routine will give values for the linear parameters a_0, a_1, s_{a0}, and s_{a1}. Back transformations are then needed to get the exponential parameters:

$$\alpha = \exp(a_0) \qquad \beta = a_1$$
$$s_\alpha^2 = \alpha\ s_{a_0}^2 \qquad s_\beta^2 = s_{a_1}^2 \tag{3.53}$$

Thus, the exponential function of Equation (3.44) has been fit to a set of data using a linear least-squares routine.

Example of Exponential Function Fit   An example of an exponential relationship is that of the dependence of a rate constant on the activation energy of the reaction. Measurement of the rates of a reaction as a function of temperatures allows calculation of the ΔH^+ of a reaction as in this example.

The following data were reported for the temperature dependence of the rate constant k_1 for the slow step in the oxidation of Fe^{2+} by hydrogen peroxide. The equation

$$k_1 = A \exp \left[\frac{-\Delta H}{RT} \right] \tag{3.54}$$

can be fit to these data to develop values for A (the preexponential factor) and ΔH^+ (activation energy) along with their errors. Activation energy should be in units of calories per mole and T is in degrees Kelvin.

Temperature (°C)	k_1 ($1\ mol^{-1}\ s^{-1}$)
15.1	40.5 ± 2.0
18.6	48.9 ± 2.5
19.75	52.8 ± 2.5
20.0	51.0 ± 2.5
20.6	58.8 ± 3.0
23.8	64.4 ± 3.0
25.1	63.4 ± 3.0
29.1	81.2 ± 4.0
33.7	102.8 ± 5.0
35.5	102.5 ± 5.0
40.0	125.2 ± 6.0

The program to perform this fitting (EXPFIT) consists of three parts: the main program, which inputs the data and calls the subroutines where the work is done; subroutine LLS, which does the least-squares curve-fitting needed; and subroutine TRANS, which does the transformation of the data and the retransformations of the parameters.

In addition to the exponential, other functional forms can also be transformed. An example is the hyperbolic function

$$y = \frac{x}{\alpha x - \beta} \tag{3.55}$$

If the transformations $y' = 1/y$ and $x' = 1/x$ are used, then the linear equation

$$y' = \alpha - \beta x \tag{3.56}$$

results.

The hyperbolic Equation (3.55) is of importance because it is the functional form that explains the relationship between the velocity of an enzymatic reaction v and substrate concentration s.

For enzymatic reactions, the rate of reaction depends on the affinity of the substrate and enzyme. If the mechanism follows the idealized model, that is, if [substrate] $>>$ [enzyme] (where brackets indicate concentration), and if

$$E + S \rightleftharpoons ES \rightleftharpoons E + P \tag{3.57}$$

then a plot of substrate concentration versus the reaction velocity produces a plot of a rectangular hyperbola through the origin. The relationship between the velocity of a reaction v and the substrate concentration s can be expressed as the Michaelis–Menten equation:

$$v = \frac{s}{s + K_{m}} V \tag{3.58}$$

in which K_{m} and V are the Michaelis–Menten constants. Here V is the velocity of the reaction when the enzyme is completely saturated with substrate and the reaction is proceeding at the maximum rate possible, and K_{m} is the substrate concentration at one-half the maximum velocity. A plot of the Michaelis–Menten equation is given in Figure 3.5. The Michaelis–Menten equation can be rearranged into the general hyperbolic Equation (3.55) using the substitutions $\alpha = 1/V$ and $\beta = -K_{m}/V$.

The Michaelis–Menten equation can be rearranged in several different ways to produce several different linear forms, as shown in the following paragraphs.

```
      PROGRAM EXPFIT
C....
C.... EXPONENTIAL FUNCTION FIT BY TRANSFORMATION
C....
      DIMENSION W(20),X(20),Y(20),YP(20),WP(20),WS(20)
      COMMON /IOUNIT/ NINP,NOUT
      DATA NINP/5/,NOUT/6/
      WRITE (NOUT,1)
    1 FORMAT (' This is Program Expfit',/)
      WRITE (NOUT,2)
    2 FORMAT (' Enter the number of points to be fit')
      READ (NINP,*) N
      DO 5 I=1,N
      WRITE (NOUT,4) I
    4 FORMAT (' Enter Point',I3,' and weight as (X,Y,W)')
    5 READ (NINP,*) X(I),Y(I),WS(I)
      DO 7 I=1,N
    7 X(I)=1.0/(X(I)+273.16)
      DO 100 II=1,2
      IF (II.EQ.2) GO TO 11
      DO 10 I=1,N
   10 W(I)=1.0
      GO TO 31
   11 DO 20 I=1,N
   20 W(I)=1.0/WS(I)**2
   31 WRITE (NOUT,32)
   32 FORMAT (' ',/,10X,'X',12X,'Y',9X,'W',/)
      WRITE (NOUT,52) (X(I),Y(I),W(I),I=1,N)
   52 FORMAT (' ',F14.6,F11.2,F10.3)
      CALL TRANS (X,Y,YP,W,WP,N,AZ,SAZ,A1,SA1,AL,SAL,BE,SBE,-1)
      CALL LLS (W,X,YP,N,A1,AZ,SA1,SAZ,STD)
      CALL TRANS (X,Y,YP,W,WP,N,AZ,SAZ,A1,SA1,AL,SAL,BE,SBE,1)
      CALL LLS (WP,X,YP,N,A1,AZ,SA1,SAZ,STD)
      CALL TRANS (X,Y,YP,W,WP,N,AZ,SAZ,A1,SA1,AL,SAL,BE,SBE,1)
      BE=BE*1.9869
      SBE=SBE*1.9869
      WRITE (NOUT,55) AL,SAL,BE,SBE
   55 FORMAT (' ',//,' Preexp. factor:',E12.3,' +/-',E12.3,//,
     X  1X,'Delta H:',F12.2,' +/-',F12.2)
  100 CONTINUE
      STOP
      END
C-----------------------------------------------------------
      SUBROUTINE LLS (W,X,Y,N,A1,AZ,SA1,SAZ,STD)
C....
C.... LINEAR LEAST SQUARES FITTING ROUTINE WITH WEIGHTS
C....
      DIMENSION W(N),X(N),Y(N)
      WW=0.0
      WX=0.0
      WY=0.0
      WXY=0.0
      WXX=0.0
      WYY=0.0
      DO 10 I=1,N
      AW=W(I)
      AX=X(I)
      AY=Y(I)
      WW=WW+AW
      WX=WX+AW*AX
      WY=WY+AW*AY
      WXY=WXY+AW*AX*AY
      WXX=WXX+AW*AX*AX
   10 WYY=WYY+AW*AY*AY
```

```
      DENOM=WW*WXX-WX*WX
      A1=(WW*WXY-WX*WY)/DENOM
      AZ=(WXX*WY-WX*WXY)/DENOM
      VSUM=0.0
      DO 20 I=1,N
   20 VSUM = VSUM + W(I) * (Y(I)-AZ-A1*X(I))**2
      SS=VSUM/(N-2)
      STD=SQRT(SS)
      SA1=SQRT(SS*WW/DENOM)
      SAZ=SQRT(SS*WXX/DENOM)
      RETURN
      END
C------------------------------------------------------------
      SUBROUTINE TRANS (X,Y,YP,W,WP,N,AZ,SAZ,A1,SA1,AL,SAL,BE,SBE,KDUM)
C....
C.... TRANSFORMATION ROUTINE : EXPONENTIAL TO LINEAR
C....
      DIMENSION X(N),Y(N),YP(N),W(N),WP(N)
      COMMON /IOUNIT/ NINP,NOUT
      IF (KDUM) 1,1,51
    1 DO 10 I=1,N
      YP(I)=ALOG(Y(I))
   10 WP(I)=W(I)*Y(I)**2
      RETURN
   51 AL=EXP(AZ)
      SAL=SAZ*AL
      BE=A1
      SBE=SA1
      WRITE (NOUT,55) AZ,SAZ,A1,SA1,AL,SAL,BE,SBE
   55 FORMAT (' ',/,' Linear parameters',/,' A0',E11.4,' +/-',E11.4,/,
     X ' A1',E11.4,' +/-',E11.4)
      WRITE (NOUT,56) AL,SAL,BE,SBE
   56 FORMAT (' Exponential parameters',/,' AL',E11.4,' +/-',E11.4,/,
     X ' BE',E11.4,' +/-',E11.4)
      VSUM=0.0
      DO 70 I=1,N
   70 VSUM = VSUM + W(I) * (Y(I)-AL*EXP(BE*X(I)))**2
      STD=SQRT(VSUM/FLOAT(N-2))
      WRITE (NOUT,75) STD
   75 FORMAT (' Overall standard deviation',F10.3)
      RETURN
      END


      This is Program Expfit

      Enter the number of points to be fit
      11
       Enter Point  1 and weight as (X,Y,W)
      15.1 40.5 2.0
       Enter Point  2 and weight as (X,Y,W)
      18.6 48.9 2.5
       Enter Point  3 and weight as (X,Y,W)
      19.75 52.8 2.5
       Enter Point  4 and weight as (X,Y,W)
      20.0 51.0 2.5
       Enter Point  5 and weight as (X,Y,W)
      20.6 58.8 3.0
       Enter Point  6 and weight as (X,Y,W)
      23.8 64.4 3.0
       Enter Point  7 and weight as (X,Y,W)
      25.1 63.4 3.0
       Enter Point  8 and weight as (X,Y,W)
```

```
29.1 81.2 4.0
 Enter Point  9 and weight as (X,Y,W)
33.7 102.8 5.0
 Enter Point 10 and weight as (X,Y,W)
35.5 102.5 5.0
 Enter Point 11 and weight as (X,Y,W)
40.0 125.2 6.0
```

X	Y	W
0.003469	40.50	1.000
0.003427	48.90	1.000
0.003414	52.80	1.000
0.003411	51.00	1.000
0.003404	58.80	1.000
0.003367	64.40	1.000
0.003353	63.40	1.000
0.003308	81.20	1.000
0.003259	102.80	1.000
0.003240	102.50	1.000
0.003193	125.20	1.000

```
Linear parameters
A0 0.1779E+02 +/- 0.5049E+00
A1-0.4052E+04 +/- 0.1507E+03

Linear parameters
A0 0.5311E+08 +/- 0.2681E+08
A1-0.4052E+04 +/- 0.1507E+03
Exponential parameters
AL 0.5311E+08 +/- 0.2681E+08
BE-0.4052E+04 +/- 0.1507E+03
Overall standard deviation    3.009

Linear parameters
A0 0.1752E+02 +/- 0.4631E+00
A1-0.3970E+04 +/- 0.1407E+03

Linear parameters
A0 0.4060E+08 +/- 0.1880E+08
A1-0.3970E+04 +/- 0.1407E+03
Exponential parameters
AL 0.4060E+08 +/- 0.1880E+08
BE-0.3970E+04 +/- 0.1407E+03
Overall standard deviation    2.962


Preexp. factor:   0.406E+08 +/-   0.188E+08

Delta H:    -7888.88 +/-      279.58
```

X	Y	W
0.003469	40.50	0.250
0.003427	48.90	0.160
0.003414	52.80	0.160
0.003411	51.00	0.160
0.003404	58.80	0.111
0.003367	64.40	0.111
0.003353	63.40	0.111
0.003308	81.20	0.062
0.003259	102.80	0.040
0.003240	102.50	0.040
0.003193	125.20	0.028

```
Linear parameters
A0 0.1809E+02 +/- 0.6289E+00
A1-0.4141E+04 +/- 0.1852E+03

Linear parameters
A0 0.7149E+08 +/- 0.4496E+08
A1-0.4141E+04 +/- 0.1852E+03
Exponential parameters
AL 0.7149E+08 +/- 0.4496E+08
BE-0.4141E+04 +/- 0.1852E+03
Overall standard deviation      0.861

Linear parameters
A0 0.1778E+02 +/- 0.4993E+00
A1-0.4050E+04 +/- 0.1491E+03

Linear parameters
A0 0.5283E+08 +/- 0.2638E+08
A1-0.4050E+04 +/- 0.1491E+03
Exponential parameters
AL 0.5283E+08 +/- 0.2638E+08
BE-0.4050E+04 +/- 0.1491E+03
Overall standard deviation      0.845


Preexp. factor:    0.528E+08 +/-    0.264E+08

Delta H:    -8047.67 +/-      296.20
```

1. *Lineweaver–Burk.* The hyperbolic form of Equation (3.58) can be linearized by the following sequence of rearrangements: Take Equation (3.51), multiply both sides by $s + K_m$, divide by V, divide by v, and the equation becomes

$$\frac{1}{v} = \frac{1}{V} + \frac{K_m}{sV} \tag{3.59}$$

This form is given in Figure 3.6, where $1/v$ is plotted against $1/s$. This linearization has been used as a basis for a graphical method for the determina-

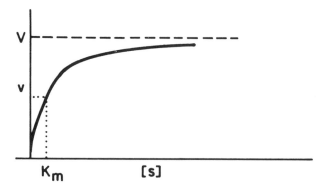

Figure 3.5 A plot of the Michaelis–Menten relationship between reaction velocity v and substrate concentration s for an enzymatic reaction.

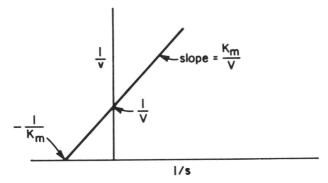

Figure 3.6 A double-reciprocal Lineweaver–Burk plot of the Michaelis–Menten relationship.

tion of the Michaelis–Menten parameters. The values for V and K_m can be read directly off the graph as shown in the figure. Garfinkel (1980) states, "Well over 90% of all published work in enzyme kinetics involves the use of this plot." However, Garfinkel goes on to say, "By now there seems to be a consensus that, of the various linearizations of the Michaelis–Menten equation that have been proposed, this is the worst." This is because the data points with small values that are inaccurate get too much weight when the reciprocals are taken.

2. *Michaelis–Menten*. Take Equation (3.58), multiply both sides by $s + K_m$, and divide both sides by vV to get

$$\frac{s}{v} = \frac{s}{V} + \frac{K_m}{V} \tag{3.60}$$

This form is given in Figure 3.7, where s/v is plotted against s. Again, the slope and intercept of the graphical plot can be used to obtain values for V and K_m.

3. *Eadie–Hofstee*. Take Equation (3.58), multiply both sides by $s + K_m$, and divide by s to get

$$v = V - \frac{vK_m}{s} \tag{3.61}$$

This form is shown in Figure 3.8, where v is plotted against v/s. Again, the values for V and K_m can be obtained from the slope and intercept of the plot.

Example of Michaelis–Menten Fit   The following is an example set of enzyme kinetic data taken from the literature (Atkinson et al. 1961).

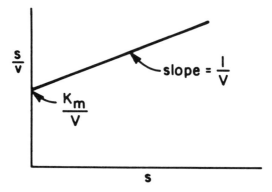

Figure 3.7 A plot of the Michaelis–Menten linearization of the Michaelis–Menten relationship.

s^a	v^b
0.138	0.148
0.220	0.171
0.291	0.234
0.560	0.324
0.766	0.390
1.460	0.493

[a]Concentration of nicotinamide mononucleotide (mM).
[b]Micromoles of nicotinamide-adenine dinucleotide formed.

Program MICMEN inputs the set of data and then evaluates the slope, intercept V, and K_m using the three linearizations of the Michaelis–Menten equation. Quite different sets of values for K_m and V are obtained from these three (nominally) identical treatments.

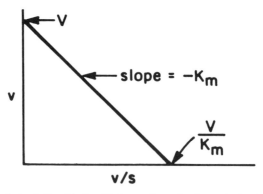

Figure 3.8 A plot of the Eadie–Hofstee linearization of the Michaelis–Menten relationship.

```
      PROGRAM MICMEN
C....
C.... MICHAELIS-MENTEN CONSTANTS FROM KINETIC DATA
C....
      DIMENSION S(20),V(20),SP(20),VP(20),W(20)
      COMMON /IOUNIT/ NINP,NOUT
      DATA NINP/5/,NOUT/6/
C....
      WRITE (NOUT,1)
    1 FORMAT (' This is Program Micmen',/)
      WRITE (NOUT,2)
    2 FORMAT (' Enter the number of points to be fit')
      READ (NINP,*) NN
      DO 10 I=1,NN
      WRITE (NOUT,3) I
    3 FORMAT (' Enter point',I2,'  (S,V)')
   10 READ (NINP,*) S(I),V(I)
      WRITE (NOUT,11) (S(I),V(I),I=1,NN)
   11 FORMAT (' ',//,8X,'S',10X,'V',/,(' ',2F11.3))
C.... LINEWEAVER-BURK METHOD
      DO 20 J=1,NN
      W(J)=1.0
      SP(J)=1.0/S(J)
   20 VP(J)=1.0/V(J)
      CALL LLS (W,SP,VP,NN,SLOPE,AINT,SA1,SAZ,STD)
      VV=1.0/AINT
      AKM=SLOPE*VV
      WRITE (NOUT,49) VV,AKM
C.... MICHAELIS-MENTEN METHOD
      DO 30 K=1,NN
      W(K)=1.0
      SP(K)=S(K)
   30 VP(K)=S(K)/V(K)
      CALL LLS (W,SP,VP,NN,SLOPE,AINT,SA1,SAZ,STD)
      VV=1.0/SLOPE
      AKM=AINT*VV
      WRITE (NOUT,49) VV,AKM
C.... EADIE-HOFSTEE METHOD
      DO 40 L=1,NN
      W(L)=1.0
      SP(L)=V(L)/S(L)
   40 VP(L)=V(L)
      CALL LLS (W,SP,VP,NN,SLOPE,AINT,SA1,SAZ,STD)
      SLOPE=-SLOPE
      WRITE (NOUT,49) AINT,SLOPE
C....
   49 FORMAT (' ',//,5X,'Michaelis-Menten constants',//,5X,'V =' ,F13.4
     1 ,/,5X,'K(M) =',F10.4)
      STOP
      END
C----------------------------------------------------------
      SUBROUTINE LLS (W,X,Y,N,A1,AZ,SA1,SAZ,STD)
C....
C.... LINEAR LEAST SQUARES FITTING ROUTINE WITH WEIGHTS
C....
      DIMENSION W(N),X(N),Y(N)
      WW=0.0
      WX=0.0
      WY=0.0
      WXY=0.0
      WXX=0.0
      WYY=0.0
      DO 10 I=1,N
      AW=W(I)
```

```
    AX=X(I)
    AY=Y(I)
    WW=WW+AW
    WX=WX+AW*AX
    WY=WY+AW*AY
    WXY=WXY+AW*AX*AY
    WXX=WXX+AW*AX*AX
10  WYY=WYY+AW*AY*AY
    DENOM=WW*WXX-WX*WX
    A1=(WW*WXY-WX*WY)/DENOM
    AZ=(WXX*WY-WX*WXY)/DENOM
    VSUM=0.0
    DO 20 I=1,N
20  VSUM = VSUM + W(I) * (Y(I)-AZ-A1*X(I))**2
    SS=VSUM/(N-2)
    STD=SQRT(SS)
    SA1=SQRT(SS*WW/DENOM)
    SAZ=SQRT(SS*WXX/DENOM)
    RETURN
    END
```

```
This is Program Micmen

 Enter the number of points to be fit
6
 Enter point 1   (S,V)
0.138 0.148
 Enter point 2   (S,V)
0.220 0.171
 Enter point 3   (S,V)
0.291 0.234
 Enter point 4   (S,V)
0.560 0.324
 Enter point 5   (S,V)
0.766 0.390
 Enter point 6   (S,V)
1.460 0.493


          S           V
        0.138       0.148
        0.220       0.171
        0.291       0.234
        0.560       0.324
        0.766       0.390
        1.460       0.493


    Michaelis-Menten constnats

    V =       0.5853
    K(M) =    0.4406


    Michaelis-Menten constnats

    V =       0.6848
    K(M) =    0.5821


    Michaelis-Menten constnats

    V =       0.6262
    K(M) =    0.4896
```

REFERENCES

Anscombe, F. J., "Graphs in Statistical Analysis," *Amr. Stat.*, **27**, 17–21 (1973).

Atkinson, M. R., J. F. Jackson, and R. K. Morton, "Nicotinamide Mononucleotide Adenyltransferase of Pig-Liver Nuclei," *Biochem. J.*, **50**, 318–323 (1961).

Bevington, P. R., *Data Reduction and Error Analysis for the Physical Sciences*, McGraw-Hill, New York, 1969.

Chaterjee, S., and B. Price, *Regression Analysis by Example*, Wiley, New York, 1977.

Cornish-Bowden, A., and R. Eisenthal, "Statistical Considerations in the Estimation of Enzyme Kinetic Parameters by the Direct Linear Plot and Other Methods," *Biochem. J.*, **139**, 721–730 (1974).

Draper, N. R., and H. Smith, *Applied Regression Analysis,* 2nd ed., Wiley, New York, 1981.

Dunford, H. B., "Equilibrium Binding and Steady-State Enzyme Kinetics," *J. Chem. Ed.*, **61**, 129–132 (1984).

Eisenthal, R., and A. Cornish-Bowden, "The Direct Linear Plot. A New Graphical Procedure for Estimating Enzyme Kinetic Parameters," *Biochem. J.*, **139**, 715–720 (1974).

Garfinkel, D., "Computer Modeling, Complex Biological Systems, and Their Simplifications," *Am. J. Physiol.*, **239**, R1–R6 (1980).

Ståhle, L. and S. Wold, "Analysis of Variance (ANOVA)," *Chemometrics and Intelligent Laboratory Systems,* **6**, 259–272 (1989).

Wilkinson, G. N., "Statistical Estimations in Enzyme Kinetics," *Biochem. J.*, **50**, 324–332 (1961).

4

MULTIPLE LINEAR REGRESSION ANALYSIS

4.1 THE BASIC METHOD

In multiple linear regression analysis, we have a set of n observations, each represented by p independent variables, $x_1, x_2, \ldots, x_p$, and a set of n dependent variables, y_i. The observation number is given the subscript i and the variable number is given the subscript j. Thus, the jth independent variable for the ith observation is denoted x_{ij} and the ith dependent variable is denoted by y_i. The mathematical model that relates the y values to the x values is assumed to be linear and of the form

$$\hat{y}_i = a_0 + a_1 x_1 + a_2 x_2 + \cdots + a_p x_p \tag{4.1}$$

where the $a_1, a_2, \ldots, a_p$ are the regression coefficients and $\hat{y}_i$ is the predicted value of the ith dependent variable. This formulation is called *multiple linear regression* because y is a linear function of x.

A specific chemical example will help to clarify the type of relationship being discussed. Given a set of organic compounds that contain only carbon, oxygen, and hydrogen atoms, the molecular weight of each compound is a multivariate relationship of these three predictors. The molecular weight (MW) is related to the three predictors in the following way:

$$MW = 1.00 \,(\#H) + 12.0 \,(\#C) + 16.0 \,(\#O)$$

where #C, #H, and #O represent number of carbon, hydrogen, and oxygen atoms. Suppose that a set of data was provided that included the values for the numbers of carbon, hydrogen, and oxygen atoms as the independent variables and the molecular weight as the dependent variable. If this data were fit by multiple linear regression analysis, the coefficients obtained would be the atomic weights of the three elements.

In a general multiple linear regression fit, the residual is defined as the difference between the observed and predicted values for the ith observation:

$$e_i = y_i - \hat{y}_i \tag{4.2}$$

It is assumed that the residuals are normally distributed, random variables. That is, the residuals contain no useful information about the relationships that might exist between the y values and the x values. Thus, they should be uncorrelated with both x and y and with each other as well.

In multiple linear regression, the values of the coefficients are found by the same general procedure that has been discussed earlier with reference to simple linear regression. The sum of the squared residuals, the SSE,

$$Q = \sum_{i=1}^{n} e_i^2 \tag{4.3}$$

is minimized by using calculus.

The best estimates of the coefficients, which are those values that minimize Q, are the solutions of the following set of normal equations:

$$
\begin{aligned}
S_{11}a_1 + S_{12}a_2 + \cdots + S_{1p}a_p &= S_{y1} \\
S_{21}a_1 + S_{22}a_2 + \cdots + S_{2p}a_p &= S_{y2} \\
&\vdots \\
S_{p1}a_1 + S_{p2}a_2 + \cdots + S_{pp}a_p &= S_{yp}
\end{aligned}
\tag{4.4}
$$

where the summations within the equations are defined as

$$S_{ij} = \sum_{k=1}^{n} (x_{ik} - \bar{x}_i)(x_{jk} - \bar{x}_j) \qquad i,j = 1,2, \ldots, p \tag{4.5}$$

$$S_{yi} = \sum_{k=1}^{n} (y_k - \bar{y})(x_{ik} - \bar{x}_i) \qquad i = 1,2, \ldots, p \tag{4.6}$$

$$\bar{x}_i = \frac{1}{n} \sum_{k=1}^{n} x_{ik} \tag{4.7}$$

$$\bar{y} = \frac{1}{n} \sum_{k=1}^{n} y_k \tag{4.8}$$

and the constant coefficient is calculated as

$$a_0 = \bar{y} - a_1\bar{x}_1 - a_2\bar{x}_2 - \cdots - a_p\bar{x}_p \tag{4.9}$$

The normal equations can be solved by the usual methods for solving systems of linear equations, and this operation is not relevant here. It is covered in the chapter on linear equations. The overall standard deviation of the fit is given by

$$S^2 = \frac{Q}{n - p - 1} \tag{4.10}$$

where Q is the sum-squared error (SSE) defined above. The coefficients, the a_i values, hae standard deviations given by

$$S_{a_j} = (S^2 c_{jj})^{1/2} \tag{4.11}$$

where c_{jj} is the jth element on the main diagonal of the inverse of the corrected sum of squares and products matrix. In matrix notation, the matrix containing the independent variables X and the matrix containing the dependent variables Y are

$$\mathbf{X} = \begin{pmatrix} x_{01} & x_{11} & \cdots & x_{p1} \\ x_{02} & x_{12} & \cdots & x_{p2} \\ x_{0n} & x_{1n} & \cdots & x_{pn} \end{pmatrix} \qquad \mathbf{Y} = \begin{pmatrix} y_1 \\ y_2 \\ \vdots \\ y_n \end{pmatrix} \tag{4.12}$$

Then, if the coefficients to be found are

$$\mathbf{a} = \begin{pmatrix} a_1 \\ a_2 \\ \vdots \\ a_p \end{pmatrix} \tag{4.13}$$

the least-squares estimates of the coefficients can be expressed as

$$\mathbf{a} = (\mathbf{X'X})^{-1} \mathbf{X' Y} \tag{4.14}$$

Then if the matrix $(\mathbf{X}'\mathbf{X})^{-1}$ is given the name $\mathbf{C}$, its main diagonal elements, c_{jj}, appear in Equation (4.14)

Analysis of Residuals

As with linear regression, multiple linear regression can be analyzed via analysis of variance. The ANOVA table for multiple linear regression is as follows:

Source of Variation	Degrees of Freedom	Sum of Squares
Due to regression	$p-2$	SSR
About regression	$n-p-1$	SSE
Total	$n-1$	SST

where n is the number of observations and p is the number of parameters (the number of independent variables plus one). As previously, the definitions of the sum square terms are as follows:

$$\text{SST} = \Sigma(\hat{y}_i - \bar{y})^2 \qquad (4.15)$$

$$\text{SSR} = \Sigma(\hat{y}_i - \bar{y})^2 \qquad (4.16)$$

$$\text{SSE} = \Sigma(y_i - \hat{y}_i)^2 \qquad (4.17)$$

$$\text{SST} = \text{SSR} + \text{SSE} \qquad (4.18)$$

The mean-square error is defined as

$$\text{MSE} = \frac{\text{SSE}}{n - p - 1} \qquad (4.19)$$

and the standard error is the square root of the MSE

$$s = (\text{MSE})^{1/2} \qquad (4.20)$$

The multiple correlation coefficient R is a measure of the adequacy of the fit to the data. It is defined as

$$R^2 = \frac{\text{SSR}}{\text{SST}} \qquad (4.21)$$

It has values between 0 and 1. Values of R^2 near 1 occur when y_i and $\hat{y}_i$ are nearly identical, which means that the model can reproduce the y_i values well. Values of R^2 near zero mean that the model is not a good one and that the y_i values are not well produced.

Reduced Model

The model of Equation (4.1) is the full model because it uses all the available independent variables. However, if only a subset of the independent variables are used to construct a model, the result will be called a reduced model. Any reduced model can be compared to the full model in a formal way as follows.

The symbol $\hat{y}_i$ represents the values predicted for y_i by the full model. Let $\hat{y}_i^*$ be the value of y_i predicted by the reduced model. The sum-squared error of the full model is

$$\text{SSE}_{\text{full}} = \sum (y_i - \hat{y}_i)^2 \tag{4.22}$$

and the SSE for the reduced model is

$$\text{SSE}_{\text{red}} = \sum (y_i - \hat{y}_i^*)^2 \tag{4.23}$$

While the full model contains p parameters, the reduced model has q parameters. The appropriate test to determine whether the reduced model represents the data as well as does the full model is to compare the SSEs. We must correct for the fact that different numbers of parameters are involved, so we compare

$$\frac{\text{SSE}_{\text{red}} - \text{SSE}_{\text{full}}}{p + 1 - q} \tag{4.24}$$

and

$$\frac{\text{SSE}_{\text{full}}}{n - p - 1} \tag{4.25}$$

The ratio of these quantities is given by

$$F = \frac{(\text{SSE}_{\text{red}} - \text{SSE}_{\text{full}})\,(n - p - 1)}{(p + 1 - q)(\text{SSE}_{\text{full}})} \tag{4.26}$$

This ratio has the F distribution with $p + 1 - q$ and $n - p - 1$ degrees of freedom.

The F value can be computed for the full model and any reduced model, and the value of F obtained can be compared to the F values in tables to determine whether the reduced model represents the data as well as the full model. This is the fundamental method used by multiple linear regression analysis routines to compare two models.

While the value of R^2 is a good guide to the quality of a model, other questions about a model should also be raised, such as

The basis of the Simpson rule approach to numerical integration is approximating the actual function to be integrated, $f(x)$, with a parabola over each narrow subinterval. In general, this will lead to errors in the estimates because the parabola will not exactly mimic the shape of the function $f(x)$. The error cannot be computed exactly (or we could add it to the approximation to obtain the exact integral value), but it can be estimated as follows. The error of estimating the area within a pair of subintervals as expressed by Equation (5.12) will govern the overall error of the procedure, so the derivation will be done on that basis. The error is given by the difference between the actual integral and the Simpson rule approximation:

$$\int_{x_1}^{x_3} f(x)\,dx - \tfrac{1}{3} h(y_1 + 4y_1 + y_3)$$

(5.14)

We assume that our function $f(x)$ can be expanded as a Taylor series about the midpoint (x_2, y_2), to get

$$f(x) = y_2 + xy_2' + \frac{x^2}{2!} y_2'' + \frac{x^3}{3!} y_2''' + \cdots$$

(5.15)

Now $y_1 = f(x_1)$ and $y_3 = f(x_3)$. The first three terms of the Taylor expansion must be zero because they constitute a parabola themselves. Therefore, the first term of Equation (5.15) that we would expect to be positive would be the x^3 term. However, when we substitute $(x^3/3!)y_2'''$ into Equation (5.14) and solve, error$_3$ is zero because of the odd symmetry. The next-higher-order term in the Taylor series is $(x^4/4!)y_2^{iv}$. When this is substituted into Equation (5.14), we obtain the following expression:

$$\text{Error}_4 \cong f^{iv}(x)h^5$$

(5.16)

after some algebraic manipulations. The absolute magnitude of this term is not so important as its functional form. It says that the error expected from a Simpson rule numerical integration—if the Taylor series well approximates the function about the midpoint of the subinterval—goes as the fourth derivative of the function and as the fifth power of the interval size h. Thus, integrals of functions for which the fourth derivative is small are approximated with little error. But functions with sizable fourth derivatives—for example, sinusoidal or exponential functions—are not approximated as well. The equation also shows that halving the interval size should decrease the expected error by a factor of 2^{-5}, or $\frac{1}{32}$. This is quite favorable because halving the interval size only doubles the computational effort but yields $\frac{1}{32}$ the expected error. These error properties of the Simpson rule approach to numerical integration account for its widespread use.

Program NISR

Simpson's rule numerical integration can be implemented in a short, straight-forward FORTRAN program, NISR, which implements Simpson's rule in an interactive routine. It integrates the equation

$$\int e^{-x} x^{a-1} \, dx \qquad (5.17)$$

for an a value and integration limits input by the user. The function itself is in the form of a statement function using the dummy variable ZZ. To integrate a different function using this program would require changing only the statement function. On execution, the routine prompts the user to input a value for the parameter a, values for the lower and upper limits of integration, the number of subintervals to be used to start the integration, and the relative accuracy desired for a stopping criterion for the iteration. The routine calculates the integral values using the Simpson rule equation, checks the relative accuracy, doubles the number of subintervals, prints a status report for the iteration, and starts over. It continues this iterative procedure until the following stopping criterion is satisfied:

$$\frac{|A_{current} - A_{last}|}{A_{current}} < \varepsilon \qquad (5.18)$$

where $A_{current}$ is the current estimate for the integral, A_{last} is the prior estimate, and ε is the error criterion. If the error criterion is unrealistically small, the routine will not terminate as desired but will lose precision due to roundoff errors as the number of subintervals becomes very large.

Equation (5.17) is very closely related to the gamma function, which is given in the following equation:

$$\Gamma(a) = \int_0^\infty e^{-x} x^{a-1} \, dx \qquad (5.19)$$

The gamma function is the extension of the factorial function to nonintegral arguments. The upper limit of integration of the gamma function is infinite, so we must approximate this upper limit while using Simpson's rule. However, this approximation will not lead to unacceptable errors because the function being integrated is dominated by the term e^{-x}, which becomes very small as x becomes large. Thus, if we use a sufficiently large value for the upper limit of integration, we will not be neglecting a significant amount of area under the curve.

The accompanying figures show two executions of program NISR using upper limits of integration of 25 and 50. An ε value of 0.001 was used in each case. Identical values for the integral were obtained, as the output shows. The

of larger size can be evaluated by reduction using cofactors to a series of 3×3 determinants. Algorithms can be devised that allow evaluation of $n \times n$ determinants with approximately $2n!$ multiplications. This factor grows so rapidly as a function of n that this approach for the evaluation of determinants is impractical. Evaluation of one 20×20 determinant would require on the order of 10^{18} operations.

There are alternative ways to evaluate determinants that are much more efficient than the brute-force method employing Cramer's rule described above. Elementary row operations can be used to make a determinant triangular, and then it can be evaluated easily. This sequence of operations can be implemented in an algorithm where the number of multiplications goes as $n^3/3$. To evaluate a 20×20 determinant by this approach requires only a few thousand operations; however, Cramer's rule is not used because superior, more efficient alternatives exist.

The Gauss–Seidel Method

Linear equation systems can be solved using an iterative approach that attempts to converge to a solution by making incremental improvements in the x values. One such method is known as the Gauss–Seidel method. Consider a 3×3 example:

$$a_{11}x_1 + a_{12}x_2 + a_{13}x_3 = b_1$$
$$a_{21}x_1 + a_{22}x_2 + a_{23}x_3 = b_2 \qquad (7.10)$$
$$a_{31}x_1 + a_{32}x_2 + a_{33}x_3 = b_3$$

These equations can be arranged to the form

$$x_1 = \frac{1}{a_{11}}(b_1 - a_{12}x_2 - a_{13}x_3)$$
$$x_2 = \frac{1}{a_{22}}(b - a_{21}x_1 - a_{23}x_3) \qquad (7.11)$$
$$x_3 = \frac{1}{a_{33}}(b_3 - a_{31}x_1 - a_{32}x_2)$$

To use this method, estimate initial values for x_2 and x_3 and plug them into the equation for x_1. The value for x_1 and the initial estimate for x_3 are plugged into the equation for x_2. The values for x_1 and x_2 are plugged into the third equation. The three steps are repeated again and again, until sufficient accuracy is obtained. Each trip through the cycle takes n^2 multiplications. In general, for an $n \times n$ set of equations, the mth approximation for the ith variable is given by

$$x_i^m = \frac{1}{a_{ii}}\left[b_i - \sum_{j=1}^{i-1} a_{ij}x_j^m - \sum_{j=i+1}^{n} a_{ij}x_j^{m-1} \right] \qquad (7.12)$$

A drawback to this method is that it is not guaranteed to converge to the correct solution. However, if the coefficient matrix is diagonally dominant, then the method is guaranteed to converge. A diagonally dominant matrix is one for which

$$a_{ii} \geq \sum_{\substack{j=1 \\ j \neq i}}^{n} |a_{ij}| \qquad (7.13)$$

That is, the diagonal element in each row of the matrix must be greater than the sum of the absolute values of all other elements in that row. In some cases, diagonal dominance can be achieved by rearranging the set of equations to place the largest elements on the main diagonal.

Gauss–Jordan Elimination Method

Linear equation systems can be solved by a direct method that systematically converts the given coefficient matrix into a form that is more easily solved. The conversion is accomplished by repeated use of elementary row operations to the coefficient matrix.

To understand the basis of the Gauss–Jordan method, it is necessary to introduce some additional matrix algebra. Recall the definition and properties of the inverse of a matrix described earlier in this section. The method for finding the inverse of a given matrix was not mentioned. An effective way to perform this operation is as follows. Starting with the coefficient matrix $\mathbf{A}$ and the identity matrix $\mathbf{I}$, perform elementary row operations on $\mathbf{A}$ to transform $\mathbf{A}$ into $\mathbf{I}$. As each operation is done to the matrix $\mathbf{A}$, do the same operation to $\mathbf{I}$ as well. On completion, $\mathbf{A}$ has become transformed into $\mathbf{I}$ and $\mathbf{I}$ has become transformed into $\mathbf{A}^{-1}$. The elementary row operations are

1. Interchange any two rows.
2. Multiply the elements of a row by a nonzero constant.
3. Add a multiple of any row to another row.

Once $\mathbf{A}^{-1}$ is obtained by this method, the values of the unknowns can be obtained by the simple matrix multiplication operation

$$\mathbf{X} = \mathbf{A}^{-1}\mathbf{B} \qquad (7.14)$$

It turns out that all the necessary operations can be carried out at once, and the method implementing this is known as the *Gauss–Jordan method*. This

theoretical chemistry. Such simulations are an important complement to experimental studies, which together constitute an active research field known as *molecular dynamics* (Levine and Bernstein 1974). One of the most powerful simulation procedures is the Monte Carlo classical trajectory method (Bunker 1974). Given the potential energy surface that describes the interactions between the atoms and molecules, one can obtain a complete dynamical microscopic picture of the collision by obtaining the numerical solution of the classical equations of motion. This solution, called a *trajectory,* gives the coordinates, momenta, and energies of all reactant, intermediate, and product species as a function of time during the collision event.

There are four principal steps in a Monte Carlo classical trajectory study:

1. Choosing the potential-energy surface for the collision event under study.
2. Selecting initial conditions for the collision partners with a Monte Carlo method so that distributions for the initial conditions represent those for the experimental study with which the trajectory results will be compared.
3. Numerically integrating the classical equations of motion.
4. Transforming the final coordinates and momenta of the trajectories to such properties as angular-momentum bond lengths and bond energies for final analysis of the trajectory results.

The potential-energy surface is usually represented by an analytic function that has a set of parameters. Values are chosen for the parameters by fitting experimental data such as vibrational frequencies, bond energies, and activation energies and by fitting theoretical ab initio electronic structure calculations. The ab initio calculations provide detailed information about the shape of the potential-energy surface far from equilibrium geometries. An example of an analytic potential-energy surface is that for the reaction

$$H + C_2H_4 \rightarrow CH_5$$

(Hase et al. 1978).

The detailed trajectory obtained from just one computational run is not sufficient for comparison with experiment. For comparison with experimental results, initial conditions are chosen by a Monte Carlo method to represent the experimental conditions under investigation. The Monte Carlo sampling procedures are as diverse as the different types of experimental conditions. For example, the reactants may be in either specific vibrational–rotational states or have a Boltzmann distribution of vibrational and rotational states specified by a temperature T. Normally, the reactants are randomly oriented; however, for some situations it is possible to align the reactants. It is clear that a different Monte Carlo sampling procedure is required for each of these cases.

Standard numerical algorithms are available for integrating the classical equations of motion. Some of the more popular are the Runge–Kutta, Adams–Moulton, and Gear algorithms. In integrating the classical equations of motion, care must be taken to use a sufficiently small time step so that accurate results are obtained. For some collision events, it may be desirable to use an integration algorithm that allows the use of a variable time step. Final trajectory properties such as angular momentum and bond lengths are obtained from the coordinates and moments with standard equations.

To illustrate the Monte Carlo classical trajectory method, details are given of a classical trajectory study by Date et al. (1984) of the energy-transfer process that occurs when a highly vibrational excited methane molecule collides with an argon atom. The study is performed using a general Monte Carlo classical trajectory computer program named MERCURY, which is available from the Quantum Chemistry Program Exchange as Program No. 453. In this simulation, methane molecules are impinged on by argon atoms. The methane molecule contains 100 kcal/mol of vibrational–rotational energy, which constitutes a large amount of internal vibrational excitation. The initial relative translational energy of the system is 5 kcal/mol. Thus, the total energy of the system is 105 kcal/mol. The total energy can be partitioned as

$$E = T + V \tag{8.13}$$

where E is the total energy, T is the kinetic energy, and V is the potential energy. The potential energy can be written as

$$V = V_{\text{inter}} + V_{\text{intra}} \tag{8.14}$$

where V_{inter} is the argon–methane (Ar–CH_4) intermolecular potential and V_{intra} is the methane intramolecular potential. There are two constants of the motion of the system: the total energy E_t and the total angular momentum J. The total angular momentum is the vector sum $J = j + l$, where j is the methane rotational angular momentum and l is the Ar–CH_4 orbital angular momentum.

The intramolecular potential of the methane molecule is represented by four Morse functions (one for each of the four carbon–hydrogen bonds) and five harmonic H—C—H bends. The Morse function has the form

$$V(r) = D\{1 - \exp[-\beta(r - r_0)]\} \tag{8.15}$$

where the parameters are as follows: r_0 is the internuclear distance for which the energy is minimum, D is the well depth or the dissociation energy, and β is a spectroscopic parameter related to the C—H stretching force constant. The harmonic bends are represented by the functional form

$$V(\theta) = f_\theta \, (\theta - \theta_0)^2 \tag{8.16}$$

where θ is the angle of an H—C—H bend, θ_0 is the angle of least strain, and f_θ is the bending force constant.

The Ar–CH$_4$ intermolecular potential is represented as a sum of five Lennard–Jones 6–12 potentials (one for the Ar–C interaction and four for the Ar–H interactions) of the functional form

$$V(r) = 4\epsilon \left[\left(\frac{\sigma}{r} \right)^{12} - \left(\frac{\sigma}{r} \right)^{6} \right] \tag{8.17}$$

where ϵ is the depth of the potential well, r is the internuclear distance, and σ is the value of r at which the potential is zero. The Lennard–Jones potential combines an attractive potential with an r^{-6} dependence with a very steeply rising repulsive potential with an r^{-12} dependence.

A Monte Carlo method is used to choose the initial conditions for the trajectories. The vibrations and rotations of methane are assigned random phases with a total energy of 100 kcal/mol. Methane is randomly oriented with respect to the argon by assignment of randomly chosen Euler angles. The collision has a fixed initial impact parameter b of 1 Å, as shown in Figure 8.1, and an initial relative translational energy of 5 kcal/mol. The impact parameter is the distance of closest approach of the centers of mass of the argon and methane in the absence of any interaction.

Once the initial conditions are chosen, the classical equations of motion are integrated. The numerical integration begins with a fourth-order Runge–Kutta method and, after six cycles, switches to a faster sixth-order Adams-Moulton predictor–corrector algorithm. As the simulation proceeds, the energy E_{vr}, l, and j are calculated after each iteration cycle. The quantity E_{vr} is the vibrational–rotational energy of the methane,

$$E_{vr} = T_{intra} + V_{intra} \tag{8.18}$$

(T_{intra} is in the center-of-mass frame), l is the Ar–CH$_4$ orbital angular momentum, and j is the methane rotational angular momentum. The orbital angular momentum is given by

$$\mathbf{l} = \mathbf{r}_{rel} \times \mathbf{p}_{rel} \tag{8.19}$$

where $\mathbf{r}_{rel}$ is the vector of position for the centers of mass of the argon and methane and $\mathbf{p}_{rel}$ is the momentum vector for the Ar–CH$_4$ relative motion. Then

$$\mathbf{j} = \sum_{i=1}^{s} \mathbf{r}_i \times \mathbf{p}_i \tag{8.20}$$

where the $\mathbf{r}_i$ and $\mathbf{p}_i$ are vectors for the five atoms of methane with respect to the methane center of mass.

The values of the methane vibrational–rotational internal energy, methane rotational angular momentum, and Ar–CH$_4$ orbital angular momentum are plotted as a function of time for a large number of trajectories. The time

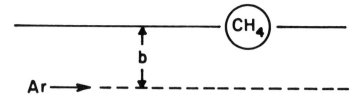

Figure 8.1   Geometry of the collision of argon with methane.

dependence of these dynamical variables gives one an intimate picture of the collision event. A typical plot is shown in Figure 8.2. The argon atom and the methane molecule interact strongly for approximately 0.1 ps. The solid line is E_{vr}, and it shows that the changes in the methane molecule's energy during the collision is much larger than the difference between the initial and final values. Strong couplings between the intramolecular motion of the CH_4 and the Ar + CH_4 intermolecular motion are evident from the modulations in the plots of E_{vr}, **I**, and **j** versus time. The plot in Figure 8.2, and many others for different initial conditions, show no evidence for long-lived collision complexes. Each trajectory is found to be characterized by only one inner turning point in the Ar + CH_4 relative motion. From studies like this one, the efficiency of energy transfer from the highly excited molecule to the less excited species is found to depend on the structure and intramolecular properties of both species. Figure 8.2 shows only one example of the many different chemical processes that can be simulated by Monte Carlo classical trajectory calculations. Classical trajectories have also provided an in-depth microscopic understanding of bimolecular and unimolecular reactions, intramolecular vibrational energy redistribution, gas–solid surface interactions, and many other phenomena. The Monte Carlo classical trajectory method is expected to become even more important for studying molecular reaction dynamics as computer technology advances.

Classical Dynamics: Scattering of Ions off Surfaces

An effective way to probe surfaces is to bombard the surface with ions and observe the scattered ions and ejected ions, neutrals, and clusters. When the ejected ions are detected, these experiments are called secondary-ion mass spectrometry (SIMS). SIMS experiments can be simulated using classical mechanics to advance understanding of the experimental observations of SIMS in the following way (Garrison and Winograd 1982).

To perform such a study, a model system must be set up. It typically consists of a representation of a segment of metal surface by an array of atoms that is four or five layers deep and on the order of 100 atoms per layer. Different crystal structures or faces of single crystals can be simulated as desired. The interactions between the individual atoms are represented by a suitable potential field. The surface of the system is bombarded with an ion with a few kiloelectronvolts of energy. Then, for each impact, one uses classi-

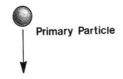

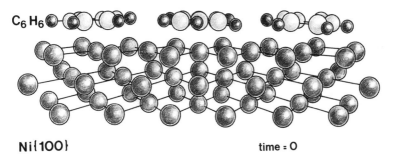

Figure 8.3 Arrangement of Ni(001) surface before Ar^+ impact. [Reprinted with permission from *Secondary Ion Mass Spectrometry, SIMS IV*, Springer-Verlag, New York, 1984. Copyright 1984, Springer-Verlag.]

simulation, has been studied (Phillips et al. 1981, Wright 1981). This method proceeds by setting up a series of mechanisms for the basic chemical processes involved, each with associated probabilities of occurrence. Random numbers are used to decide which events occur at any given time. Large numbers of events occur, and the simulation collects the results as statistical averages.

In one such study (Phillips et al. 1981), gas–solid chromatography was simulated. The molecules flow in a carrier-gas stream over a solid stationary phase. Individual molecules were put through the chromatographic column one at a time. This precludes taking interactions between molecules into account, but it is a reasonable approximation for the dilute concentrations characteristic of the mobile gas phase. Peak shapes were obtained for comparison to experimentally observed chromatographic peaks.

In another study (Wright 1981), the high-performance liquid chromatography (HPLC) column was simulated. Diffusion of the molecules of interest in the mobile phase was neglected as unimportant. The model was focused on describing the surface composition of the stationary phase and the interactions between it and the molecules of interest. The adsorption–desorption events were modeled by a two-step retention mechanism. It was assumed that there was no competition between molecules for adsorption sites, that is, infinite dilution. A molecule in the mobile phase absorbed on an interaction site with some probability. Once there, it could either return directly to the

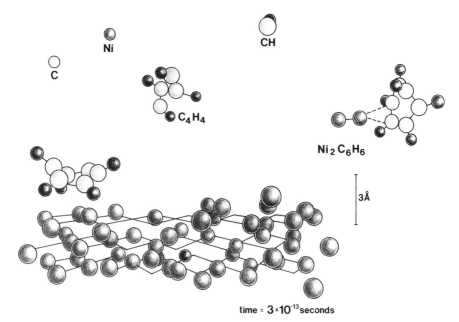

Figure 8.4   Arrangement of Ni(001) surface and ejected particles 3×10^{-13} s after Ar^+ impact. [Reprinted with permission from *Secondary Ion Mass Spectrometry, SIMS IV*, Springer-Verlag, New York, 1984. Copyright 1984, Springer-Verlag.]

mobile phase or it could jump to a second, nearby adsorption site. Eventually, it returned to the mobile phase. The model also provided for injecting molecules into the column, moving them through the column while in the mobile phase, and detecting them on elution.

The adjustable parameters that were accessible in the model were as follows:

1. Probability of a molecule encountering an interaction site on the stationary phase
2. Probability of a molecule moving from the first interaction site to the second
3. Probability of an adsorbed molecule returning to the mobile phase
4. Sample size and number of interaction sites of each type on the stationary phase

A series of HPLC experimental studies was done for comparison. Benzene and aniline were put through three stationary phases with differing compositions. Spherisorb was the most polar surface because it had the most silanol groups, Lichrosorb had an intermediate number of silanol groups, and μ-Bondapak C-18 had the least polar surface.

The discrete-event simulation was used to compare the calculated versus observed peaks of benzene and aniline as a function of stationary-phase polarity. The authors found that their two-site model could produce simulated peaks that could be compared to the experimental ones for analysis. Thus, they showed that HPLC retention could be explained satisfactorily by the two-step retention model. The simulation was an integral part of the overall project, allowing the direct comparison of calculation and observation.

Quadrupole Mass Analyzer Simulation

The quadrupole mass filter (QMF) is a very widely used type of mass spectrometer, and it is incorporated into many commercial instruments. A QMF can scan quickly [1 amu (atomic mass unit)/ms], and this has made it attractive for incorporation into gas chromatograph–mass spectrometer instruments (GC-MS). However, its characteristics are not fully understood even though they have been extensively studied (e.g., Campana 1980). Computer software simulation of the QMF allows investigation of characteristics, such as individual ion trajectories, that are not experimentally observable (Campana and Jurs 1980).

An ideal QMF consists of four long hyperbolic cylinders in a square array with the inside radius of the array equal to the smallest radius of curvature of the hyperbolas. In practice, this geometry is approximated by four parallel cylindrical rods mounted at the corners of a square at a distance of $2r_0$ from the opposite rod. Opposite rods are electrically connected. If the radius of the rods is made $1.1486r_0$, then the ideal electric field that would be provided by the hyperbolic rods is well approximated by the circular rods. Voltage V is the peak RF voltage, ωt are impressed between opposite pairs of rods. Voltage V is the peak RF voltage, ω is the angular frequency $2\pi f$, and t is time. The field potential in the quadrupole field is given by

$$\Phi(x,y,z,t) = \frac{(U + V \cos \omega t)(x^2 - y^2)}{r_0^2} \tag{8.31}$$

where x, y, and z are the Cartesian coordinates within the field. Since there is no applied potential in the z direction, there is no z term in the field potential. This field potential results in the following electric fields in the x and y directions:

$$E_x = -\frac{2(U + V \cos \omega t)x}{r_0^2} \tag{8.32}$$

$$E_y = +\frac{2(U + V \cos \omega t)y}{r_0^2} \tag{8.33}$$

For a certain set of quadrupole operating conditions, a particular ion (characterized by its mass, energy, entrance location, etc.) may have a stable trajectory and survive transit through the QMF. The ions are injected into the field of the QMF parallel to the z axis with an energy determined by the accelerating potential they have fallen through. The ions proceed with a constant velocity in the z direction, and they undergo complicated oscillatory motions in the x and y directions. For a particular set of operating conditions, only ions of a particular m/z will move through the mass analyzer to be collected at the detector. All other ions will collide with the rods and be lost.

The parameters that characterize the analyzer are U, V, $f = \omega/2\pi$, and r_0. Mass scanning is accomplished by varying U and V, keeping their ratio constant, while keeping f constant.

From $F = ma$, the equations of motion of an ion of mass m in the electric fields are as follows:

$$m \frac{d^2x}{dt^2} - eE_x = 0 \tag{8.34}$$

$$m \frac{d^2y}{dt^2} + eE_y = 0 \tag{8.35}$$

$$m \frac{d^2z}{dt^2} = 0 \tag{8.36}$$

The motion of an ion in either direction is independent in an ideal field. These equations are generally used with the following variable substitutions:

$$a = \frac{4eU}{m\omega^2 r_0^2} \tag{8.37}$$

$$q = \frac{2eV}{m\omega^2 r_0^2} \tag{8.38}$$

$$\xi = \tfrac{1}{2}\omega t \tag{8.39}$$

Substitutions of these dimensionless quantities and rearrangement of the equations yields a final equation of the following form for the x-direction motion:

$$\frac{d^2x}{d\xi^2} = (a_x + 2q_x \cos 2\xi)x = 0 \tag{8.40}$$

An equation with y replacing x expresses the y-direction motion. These equations are special types of linear second-order differential equations known as *Mathieu equations*. Mathieu equations have been studied intensively and are well understood mathematically (McLachlan 1947).

For certain sets of values for a and q, there are stable solutions to Mathieu equations that correspond to stable ion trajectories through the length of the QMF. Detailed consideration of the a–q space and its region of stability are beyond the scope of this discussion but can be found in the literature (e.g., Campana 1980). Briefly, there is a region of the a–q space, bounded by two polynomials, within which the a,q values lead to stable trajectories. The ion trajectories through the QMF are helical; they can be most easily viewed by decomposing the motion into the two orthogonal components in the x–z and the y–z planes. The region of stability is independent of the initial velocity of the ion and the relative phase of the RF field at the time of entry. In practice, a values between 0 and 0.24 and q values between 0 and 0.9 can be used. The exact values of a and q and their ratio determine, to a large extent, the properties of the QMF. The resolution of the QMF ($R = m/\Delta m$) is determined by the value of the a/q ratio.

Trajectories can be found by numerical integration of the differential equations of motion.

Another approach, implemented in the program QMAS (see list at end of chapter), is to construct a computer model of the QMF. This is a simulation approach in which the electric field is expressed as a function of location and time. The electric field interacts with a charged particle to impart an acceleration that causes a displacement of the ion. The path of an ion can be mapped as it passes through the QMF by repetitive iterations of the cycle described above. The results can be viewed on a graphics display terminal.

A program nearly identical to QMAS was used to generate some example ion trajectories reported in a paper by Campana (1980). The QMF parameters used in these simulations were as follows: inscribed radius r_0, 0.277 cm; field length l, 15.24 cm; mass resolution R, 100; RF frequency f, 2.5 MHz; a parameter, 0.2344; q parameter, 0.7037; RF peak voltage V, 690.38 V; DC voltage U, 114.99 V. The ions injected into the field were of mass 100 amu. The x and y entrance coordinates were set to 0.01 cm, that is, the ion entered just off center to the main axis of the QMF. The x and y velocity components for the ions were 47.053 m/s. The z velocity component was 3105.496 m/s. The initial kinetic energy of the ions was 5.0 eV. For these conditions, the transit time through the QMF was 4.907×10^{-5} s. The RF field had a phase angle of zero at the time the ions entered the QMF. During the time the ion was in transit through the field, 122.69 RF cycles occurred. The 100-amu ion was considered to have been formed at a point source 0.660 cm in front of the entrance to the QMF with an initial total energy of 5 eV.

Several different sets of a–q values were chosen, and a trajectory was computed for each. Figures 8.5–8.7 show the trajectories resulting from the following sets of q,a values: (0.7002, 0.2333), (0.7037, 0.2344), (0.7072, 0.2356). Different sets of a and q result in trajectories that are relatively stable, as in Figure 8.6, or relatively unstable, as in Figures 8.5 and 8.7. Figure 8.8 shows the trajectory of an ion of mass 99 amu when injected into the QMF tuned to pass only 100-amu ions. Figure 8.9 shows the trajectory of an ion of mass 101 amu when injected into the same QMF. Since the QMF was tuned to pass ions of mass 100 amu and has a resolution of 100 (unit resolution at mass 100), the ions with masses of 99 and 101 amu cannot pass all the way through the QMF.

In addition to plotting the individual trajectories of ions passing through the QMF, many additional simulation experiments can be performed. An example is the study of the average properties of ions as they pass through the

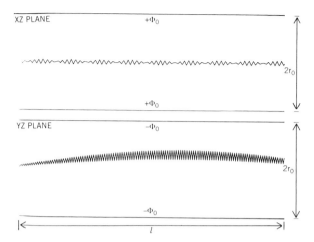

Figure 8.5 Ion trajectory of the $m/z = 100$ ion calculated for a and q values near the boundary of the a–q stability diagram. [Reprinted with permission from *International Journal of Mass Spectrometry and Ion Physics*, **33**, 101 (1980). Copyright 1980, Elsevier Science Publishers.]

QMF. The parameters that characterize the interaction between the ion and the electric field within the QMF can be averaged to approximate the behavior of ions in a real laboratory experiment. Such a set of experiments was reported by Campana and Jurs (1980). The full details of the work can be obtained from the original paper. Briefly, the experiment done was as follows. Ions were injected into the QMF in locations spread about the entrance

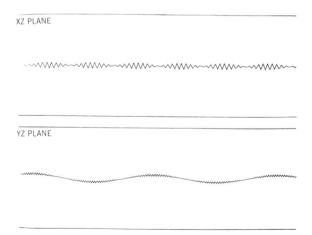

Figure 8.6 Ion trajectory of the $m/z = 100$ ion calculated for a and q values well within the a–q stability diagram. [Reprinted with permission from *International Journal of Mass Spectrometry and Ion Physics*, **33**, 101 (1980). Copyright 1980, Elsevier Science Publishers.]

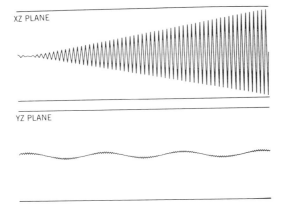

Figure 8.7   Ion trajectory of the $m/z = 100$ ion calculated for a and q values near the boundary of the a–q stability diagram. [Reprinted with permission from *International Journal of Mass Spectrometry and Ion Physics*, **33**, 101 (1980). Copyright 1980, Elsevier Science Publishers.]

aperture, with varying angles of entrance with respect to the z axis of the QMF, with varying a,q values, with varying RF phase angles, but with fixed energy (5.0 eV). Each study was done with unit resolution. In total, 2600 ions were put through the QMF per simulation. The computation of the trajectory of any ion was terminated if the ion strayed too far from the central axis of the QMF. The location of exit of each ion from the QMF was stored. The fraction of ions surviving transit through the QMF was found to fall off for higher mass ions (at unit resolution). A three-dimensional histogram showing the exit

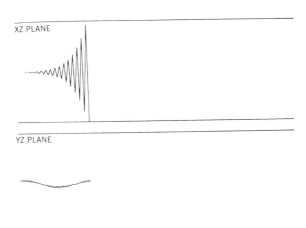

Figure 8.8   Ion trajectory of the $m/z = 99$ ion calculated at quadrupole operating conditions set to pass only the $m/z = 100$ ion. [Reprinted with permission from *International Journal of Mass Spectrometry and Ion Physics*, **33**, 101 (1980). Copyright 1980, Elsevier Science Publishers.]

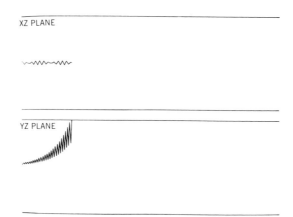

Figure 8.9 Ion trajectory of the $m/z = 101$ ion calculated at quadrupole operating conditions set to pass only the $m/z = 100$ ion. [Reprinted with permission from *International Journal of Mass Spectrometry and Ion Physics*, **33**, 101 (1980). Copyright 1980, Elsevier Science Publishers.]

distribution is given in Figure 8.10. The ions were focused toward the center of the QMF and more along the y axis than the x axis. These results were in accord with experimental observations of QMF behavior.

LIST OF VARIABLES USED IN PROGRAM QMAS

Quadrupole Mass Filter Operating Conditions

MMIN, MMAX	Minimum and maximum masses of ions to be used
RES	Resolution
RATAQ	Ratio of parameters $a–q$; scan line slope
QL,QR	Values of q where scan line intersects the left and right edges of the stability diagram
RZ	Inscribed circle radius in the quadrupole (cm)
F	Frequency of the RF field (MHz)
U	DC voltage
V	RF peak voltage
EL	Length of quadrupole rods (cm)
PSD	Distance from point source to quadrupole entrance aperture (cm)

Ion Characteristics

G	Mass of injected ion (amu)
E	Energy of injected ion (eV)
XZ,YZ	x and y entrance locations (cm)
VT	Velocity of injected ion
PHI,THETA	Entrance angles of injected ion

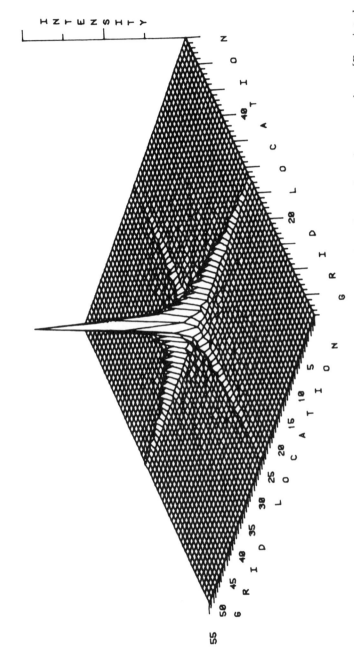

Figure 8.10 Three-dimensional histogram showing the ion exit distribution from the quadrupole mass analyzer. [Reprinted with permission from *International Journal of Mass Spectrometry and Ion Physics*, **33**, 101 (1980). Copyright 1980, Elsevier Science Publishers.]

VX, VY, VZ	Velocities in the x, y, and z directions
T	Transit time of ion through quadrupole
N	Number of evaluations of force field to be made
CNOS	Number of cycles of RF field seen by ion during transit through field
EX,EY	Electric fields in x and y directions
FX,FY	Forces in x and y directions
AX,AY	Accelerations in x and y directions
DX,DY	Displacements in x and y directions
XD,YD	Storage of positions of ion at time increments

Other Parameters

MFREK	Number of points to be plotted

```
      PROGRAM QMAS
C
C     QUADRUPOLE MASS ANALYZER SIMULATOR
C
      DIMENSION XD(1000),YD(1000)
      COMMON /IOUNIT/ NINP,NOUT
      DATA PI/3.14159/,NINP/5/,NOUT/6/
      WRITE (NOUT,209)
  209 FORMAT (' This is Program Qmas',/)
      WRITE (NOUT,219)
  219 FORMAT (' Enter the desired resolution')
      READ (NINP,*) RES
      WRITE (NOUT,229)
  229 FORMAT (' Enter minimum and maximum masses in a.m.u.')
      READ (NINP,*) MMIN,MMAX
C     CON IS THE NUMBER OF ITERATIONS PER RF CYCLE
      CON=100.0
      RZ=0.277
      F=1.8
      RATAQ=2.*(.16784-.12588/RES)
      CALL INTR (QL,QR,RATAQ)
      I=0
 1003 CALL TRAJEK(QL,QR,MMIN,MMAX,RATAQ,RES,RZ,N,A,Q,XZ,YZ,VX,VY,VZ,F,
     1TT,U,V,G,I,MFREK,CON)
      XZS=XZ
      YZS=YZ
      VXS=VX
      VYS=VY
      JNUMB=1
      NDEN=1
      IF (NDEN.EQ.0) NDEN=1
      DO 1010 JJ=1,JNUMB
      XZ=XZS
      YZ=YZS
      VX=VXS
      VY=VYS
      CALL QUAD(XD,YD,VX,VY,VZ,TT,F,U,V,RZ,G,N,XZ,YZ,R,XMAX,YMAX,
     XIPLT,JJ,A,Q,RATAQ,MFREK,NDEN)
 1010 CONTINUE
      STOP
      END
```

```
C----------------------------------------------------------------
      SUBROUTINE TRAJEK(QL,QR,MMIN,MMAX,RATAQ,RES,RZ,N,A,Q,XZ,YZ,VX,VY,
     1 VZ,F,TT,U,V,G,I,MFREK,CON)
C.... INPUT ION ENTRANCE CONDITIONS; PERHAPS SET UP QUAD TO FOCUS IT
      COMMON /IOUNIT/ NINP,NOUT
      DATA PI/3.14159/,EL/15.24/,PSD/0.660/
      WRITE (NOUT,109)
  109 FORMAT (' Enter ion mass in a.m.u.')
      READ (NINP,*) G
      WRITE (NOUT,119)
  119 FORMAT (' Enter ion energy in electron volts')
      READ (NINP,*) E
      WRITE (NOUT,129)
  129 FORMAT (' Enter off axis x and y entry locations')
    6 READ (NINP,*) XZ,YZ
      IF (XZ.LT.RZ.AND.YZ.LT.RZ) GOTO 7
      WRITE (NOUT,139)
  139 FORMAT (' Too far from center ... try again')
      GOTO 6
    7 VT=SQRT((2.0*E*1.60219E-12)/(G*1.66042E-24))
      THETA=ATAN(YZ/XZ)
      R=SQRT(XZ**2+YZ**2)
      PHI=ATAN(R/PSD)
      PHI=180.0*PHI/PI
      THETA=180.0*THETA/PI
      VV=VT*SIN(PHI*PI/180.0)
      VX=VV*COS(THETA*PI/180.0)
      VY=VV*SIN(THETA*PI/180.0)
      VZ=VT*COS(PHI*PI/180.0)
      IF (I .EQ. 3) GO TO 1
      MDIFF=MMAX-MMIN
      QDIFF=QR-QL
      QINC=QDIFF/MDIFF
      Q=QL+QINC*(G-MMIN)
      A=Q*RATAQ
      U=1.66042*A*PI**2*G*F**2*RZ**2/1.60219
      V=2.*U/RATAQ
      WRITE (NOUT,149)
  149 FORMAT (/,' Entry conditions')
      WRITE (NOUT,159) A,Q,RATAQ,V,U,RES
  159 FORMAT (' A = ',F12.5,/,' Q = ',F12.5,/,' A/Q = ',F10.5,/,
     X  ' V =',F10.2,/,' U = ',F9.2,/,' Resolution = ',F5.0)
    1 TT=EL/VZ
      N=TT*F*1.0E+06*CON
      CNOS=TT*F*1.0E+06
      MFREK=CNOS*2.0
      WRITE (NOUT,169)  CNOS,VX,VY,VZ,THETA,PHI,TT
  169 FORMAT (' No. of RF cycles = ',F7.2,/,' Entry velocities = ',
     X   3F7.0,/,' Entry angles (theta,phi) = ',2F7.1,/,
     X  ' Expected time in RF field = ',E12.4,' SEC.')
      RETURN
      END
C----------------------------------------------------------------
      SUBROUTINE QUAD(XD,YD,VX,VY,VZ,TT,F,U,V,RZ,G,N,XZ,YZ,R,XMAX,YMAX,
     1 XIPLT,JJ,A,Q,RATAQ,MFREK,NDEN)
C.... PUT AN ION THROUGH THE QUADRUPOLE MASS FILTER
      DIMENSION XD(1000), YD(1000)
      COMMON /IOUNIT/ NINP,NOUT
      DATA PI/3.14159/
      YMAX=0.0
      XMAX=0.0
      IF (JJ .NE. 1) MFREK=MFREK-1
      XM=1.0/(G*1.66042E-24)
      RZS=RZ**2
      FRAC=FLOAT(JJ-1)/FLOAT(NDEN)
```

```
      M=0
      KK=1
      XD(KK)=XZ
      YD(KK)=YZ
      NO=N/MFREK
      DT=TT/N
      TZ=FRAC/(F*1.0E+06)
      TWOPI=2.0*PI*F*1.0E+06
      ITR=0
      DO 20 I=1,N
      T=DT*(I-1)
      FAC=(U+V*COS(TWOPI*(T-TZ)))/RZS
      EX=-FAC*XZ
      EY=FAC*YZ
      FX=1.60219E-12*EX
      FY=1.60219E-12*EY
      AX=FX*XM
      AY=FY*XM
      DX=VX*DT+(AX*DT**2)/2.0
      DY=VY*DT+(AY*DT**2)/2.0
      XZ=DX+XZ
      YZ=DY+YZ
      M=M+1
      IF (M-NO) 11,8,11
    8 KK=KK+1
      XD(KK)=XZ
      YD(KK)=YZ
      M=0
   11 XTEM= ABS(XZ)
      YTEM= ABS(YZ)
      IF (XTEM.GT.XMAX) XMAX=XTEM
      IF (YTEM.GT.YMAX) YMAX=YTEM
      VX=VX+AX*DT
      VY=VY+AY*DT
      IF (XTEM.GE.RZ.OR.YTEM.GE.RZ) GO TO 24
   20 CONTINUE
      ITR=1
   24 KK=KK+1
      XD(KK)=XZ
      YD(KK)=YZ
      IPLT=KK
      R=SQRT(XZ**2+YZ**2)
      MFREK=MFREK+1
      IF (JJ.NE.1) RETURN
      THETA=ATAN(VY/VX)
      VV=SQRT(VY**2+VX**2)
      PHI=ATAN(VV/VZ)
      PHI=180.0*PHI/PI
      THETA=180.0*THETA/PI
      WRITE (NOUT,109)
  109 FORMAT (/,' Exit conditions')
      IF (ITR.EQ.0) WRITE (NOUT,119)
  119 FORMAT (' Ion not transmitted')
      IF (ITR.EQ.1) WRITE (NOUT,129) VX,VY,VZ,THETA,PHI
  129 FORMAT (' Exit velocities = ',3F10.0,/,' Exit angles ',
     X ' (theta,phi) = ',2F10.3)
      IF (ITR.EQ.1) WRITE (NOUT,139) IPLT,XZ,YZ,R,XMAX,YMAX
  139 FORMAT (' No. of points per trajectory = ',I5,/,
     X ' Exit location (X,Y,R) = ',3F10.3,/,
     X ' Max. distances from axis (X,Y) = ',2F10.3)
      RETURN
      END
C-------------------------------------------------------------
      SUBROUTINE INTR (QL,QR,RATAQ)
C     BISECTION METHOD OF  ROOT FINDING
C     FIND INTERSECTIONS OF SCAN LINE WITH EDGES OF STABILITY DIAGRAM
      DATA  ACC/1.0E-05/
      SCAN(Q,RATAQ)=Q*RATAQ
```

```
        A(Q)=(.5*Q**2)-(0.05469*Q**4)+(0.01259*Q**6)-(0.00364*Q**8)
        B(Q)=1.0-Q-(0.12500*Q**2)+(0.01563*Q**3)-(0.00065*Q**4)-
      1(.00034*Q**5)+(.00025*Q**6)
        FUNC1(Q,RATAQ)=A(Q)-SCAN(Q,RATAQ)
        FUNC2(Q,RATAQ)=B(Q)-SCAN(Q,RATAQ)
        DO 100 II=1,2
        L=II-1
        A1=0.0
        A2=0.706
        IF (L.EQ.1) A1=1.0
        I=0
        IF(L.EQ.0) GO TO 40
        B1=FUNC2(A1,RATAQ)
        B2=FUNC2(A2,RATAQ)
        GO TO 50
40      B1=FUNC1(A1,RATAQ)
        B2=FUNC1(A2,RATAQ)
50      IF (B1) 51,51,86
51      IF (B2) 86,52,52
52      X=(A1+A2)/2.
        I=I+1
        IF (L.EQ.0) GO TO 60
        Y=FUNC2(X,RATAQ)
        GO TO 80
60      Y=FUNC1(X,RATAQ)
80      IF (ABS(X-A2)-ACC) 86,86,83
83      IF (Y) 84,86,85
84      A1=X
        GO TO 52
85      A2=X
        GO TO 52
86      Y=RATAQ*X
        IF (L.EQ.0) QL=X
        IF (L.EQ.1) QR=X
100     CONTINUE
        RETURN
        END


 This is Program Qmas

 Enter the desired resolution
100
 Enter minimum and maximum masses in a.m.u.
99 101
 Enter ion mass in a.m.u.
100
 Enter ion energy in electron volts
5.
 Enter off axis x and y entry locations
0.05 0.05

 Entry conditions
 A =      0.23446
 Q =      0.70373
 A/Q =    0.33316
 V =      357.88
 U =       59.62
 Resolution =  100.
 No. of RF cycles =   88.82
 Entry velocities = 23399. 23399.308866.
 Entry angles (theta,phi) =    45.0    6.1
 Expected time in RF field =   0.4934E-04 SEC.
```

```
Exit conditions
Exit velocities =       564163.     175639.     308866.
Exit angles (theta,phi) =       17.293     62.403
No. of points per trajectory =    179
Exit location (X,Y,R) =        0.103     -0.049      0.114
Max. distances from axis (X,Y) =          0.133      0.188
```

REFERENCES

General

Bunker, D. L., "Simple Kinetic Models from Arrhenius to the Computer," *Accts. Chem. Res.*, **7**, 195–201 (1974).

Fluendy, M., "Monte Carlo Studies," in *Markov Chains and Monte Carlo Calculations in Polymer Science*, G. G. Lowry (ed.), Marcel Dekker, New York, 1970, Chapter 3.

Güell, O. A., and J. A. Holcombe, "Stochastic Chemometrics: Analytical Applications of Monte Carlo Techniques," *Anal. Chem.*, **62**, 529A (1990).

Lykos, P. G., *Computer Modelling of Matter*, American Chemical Society, Washington, DC, 1978.

Schrage, L., "A More Portable Fortran Random Number Generator," *A.CM. Trans. Math. Software*, **5**, 132 (1979).

Zelen, M., and N. Severo, "Methods of Generating Random Numbers and Their Applications," in *Handbook of Mathematical Functions with Formulas, Graphs, and Mathematical Tables*, M. Abramowitz and I. A. Stegun (eds.), National Bureau of Standards Applied Mathematics Series, No. 55, U.S. Government Printing Office, Washington, DC, 1964, Section 26.8, pp. 949–953.

Monte Carlo Integration

Press, W. H., B. P. Flannery, S. A. Teukolsky, and W. T. Vetterling, *Numerical Recipies*, Cambridge University Press, 1986, p. 221.

Water Simulation

Rahman, A., and F. H. Stillinger, "Molecular Dynamics Study of Liquid Water," *J. Chem. Phys.*, **55**, 3336 (1971).

Classical Trajectory

Date, N., W. L. Hase, and R. G. Gilbert, "Collisional Deactivation of Highly Vibrationally Excited Molecules. Dynamics of the Collision Event," *J. Chem. Phys.*, **88**, 5135–5138 (1984).

Hase, W. L., "MERCURY: A General Monte Carlo Classical Trajectory Program," *Q.C.P.E. Bull.* **3**, 453 (1983).

Hase, W. L., G. Mrowka, R. J. Brudzynski, and C. S. Sloane, "An Analytical Function Describing the H + C_2H_4 = C_2H_5 Potential Energy Surface," *J. Chem. Phys.*, **69**, 3548–3562 (1978).

Levine, R. D., and R. B. Bernstein, *Molecular Reaction Dynamics*, Oxford University Press, New York, 1974.

Surface Scattering

Garrison, B. J., "Organic Molecule Ejection from Surface Due to Heavy Particle Bombardment," *J. Am. Chem. Soc.*, **104**, 6211 (1982).

Garrison, B. J., "Mechanisms of Organic Molecule Ejection in SIMS and FABMS Experiments," in *Secondary Ion Mass Spectroscopy, SIMS IV*, A. Benninghoven, J. Okano, R. Shimizu, and H. W. Werner (eds.), Springer-Verlag, Berlin, 1984.

Garrison, B. J., and N. Winograd, "Ion Beam Spectroscopy of Solids and Surfaces," *Science*, **216**, 805 (1982).

Garrison, B. J., and N. Winograd, "Ion Beams and Lasers," *Chemtech*, **23**, 25 (1993).

Chromatographic Simulation

Phillips, J. B., N. A. Wright, and M. F. Burke, "Probabilistic Approach to Digital Simulation of Chromatographic Processes," *Sep. Sci. Technol.*, **16**, 861 (1981).

Wright, N. A., "Computer Assisted Investigations of Chromatographic Processes," master's thesis, University of Arizona, 1981.

Quadrupole Mass Filter Simulation

Campana, J. F., "Elementary Theory of the Quadrupole Mass Filter," *Internatl. J. Mass. Spectrom. Ion Phys.*, **33**, 101–117 (1980).

Campana, J. E., and P. C. Jurs, "Computer Simulation of the Quadrupole Mass Filter," *Internatl. J. Mass Spectrom. Ion Phys.*, **33**, 119–137 (1980).

Dawson, P. H., *Quadrupole Mass Spectrometry and Its Applications*, Elsevier, Amsterdam, 1976.

McLachlan, N. W., *Theory and Application of Mathieu Functions*, Oxford University Press, New York, 1947.

Reiser, H. P., R. K. Julian, Jr., and R. G. Cooks, "A Versatile Method of Simulation of the Operation of Ion Trap Mass Spectrometers," *Internatl. J. Mass Spectrom. Ion Processes*, **121**, 49–63 (1992).

9

SIMPLEX OPTIMIZATION

9.1 THE SIMPLEX METHOD

Many activities of chemists involve observing the output or response of a system, whether it is an instrument or a mathematical model or a reaction, as a function of a number of experimental variables. For example, the common activity of tuning an instrument means adjusting the instrumental settings in order to seek the best response. The best response may be the best sensitivity or the best selectivity or some combination of both. In the development of new analytic chemical methods and in the improvement or extension of existent ones, it is common to investigate the effects of experimental variables on the results. Variables such as pH, temperature, and reagent concentration must be varied to find those that impact the observed results. Then the values of these important variables are optimized to improve the results of the method. Improvement can mean increased selectivity, increased sensitivity, decreased interferences, faster operation, greater precision, or some other desired outcome.

The use of chemical instrumentation nearly always involves tuning the instrument for the best response for the experimental measurement being made. This is a classical optimization problem, especially since the instrumental settings are ordinarily interrelated. Consider a specific example, atomic absorption spectroscopy with a flame atomizer. Experimental variations that are directly related to the observed signal level and precision of measurements include the hollow-cathode lamp current, widths of the various slits in the monochromator, flame conditions, fuel and oxidant flow rates, burner position, spectral line used, integration time used for detection, and many others as well. In organic synthesis, the conditions are varied in order to seek the best yield of the desired product. All of these activities, and many more, involve optimization. A good, workable definition of optimization has appeared as follows: "the collective process of finding the best set of conditions required to achieve the best result from a given situation" (Beveridge and Schechter 1970).

The first obstacle to optimization of real systems is the multidimensional nature of the problem; that is, there are usually many variables to optimize simultaneously. If the variables that are being investigated are independent of one another, it is a straightforward task to vary them one at a time while observing the response. Independent variables can be optimized individually. However, in real situations, such complete independence is rarely found. The variables to be considered are almost always related to one another so that one must study them collectively. Multivariate methods must be used.

There are many types of optimization methods, four of which we will discuss here: single-factor variation, grid searches, random approaches, and the simplex method. The *single-factor variation* method involves systematically changing one factor at a time while holding all the remaining factors constant. It is well suited to cases where the variables being investigated are independent. However, the method is often applied for convenience to cases where the variables are interdependent even though it is not really applicable. When there are many variables, and they are correlated, then the single-factor variation method can lead to severely incorrect values of the factors.

Grid searches involve evaluation of the response for several different sets of values of the variables chosen so that they form a grid in the factor space. The first step of the optimization can use a coarse grid that covers a substantial fraction of the factor space. Once a region of the factor space that has good response values is located, the grid size can be decreased in order to focus more narrowly on the best region. Then the process can be repeated with progressively decreasing grid sizes. This method is reasonably sure to find good optima, but at a cost of many evaluations of the response.

The *random approach* involves choosing factor values randomly throughout the factor space. This ensures that the factor space is sampled far and wide. This method can be wasteful in that a great deal of time can be spent evaluating responses in uninteresting regions of the factor space. To ensure that good optima are found, it may be necessary to sample a very large number of factor values, which is costly and time-consuming.

In the fourth optimization method discussed here, the *simplex method*, random variation is replaced with an orderly, statistical design. Evaluations of the response of the system are made, and then this information is used to determine the direction to move for the next evaluation. This cycle is repeated until the optimum is found. The overall operation is driven quite efficiently by the use of all the information that is available at any given time.

The Simple Simplex Method

The simplex method is a sequential optimization method that involves repeated observation of the system response, selection of new values for the variables, followed by another observation, and so on. The method can be visualized for optimization problems involving just two variables as a method for tracking on the response surface where one axis corresponds to variable 1, a second axis corresponds to variable 2, and the third axis corresponds to the response. This corresponds to seeking hills or ridges on a topographical map. A geometric construct called a *simplex* is used as the method for tracking about on the response surface.

A *simplex* is a geometric figure with its number of vertices equal to one more than the number of dimensions of the factor space. A simplex in two dimensions is a triangle; in three dimensions it is a tetrahedron. In four or more dimensions, simplexes cannot be drawn or readily visualized, but the geometry is intact and the method is applicable. So, in general, the number of vertices of a simplex is the number of dimensions in the factor space (equal to the number of variables being investigated) plus 1. Figure 9.1 shows a simplex, with its vertices labeled *B*, *N*, and *W* superimposed on a two-dimensional factor space.

A new simplex can be constructed adjacent to an existing one by retaining all except one vertex of the existing one and creating just one new vertex. The simplex in Figure 9.1 can be converted into an adjacent one by eliminating

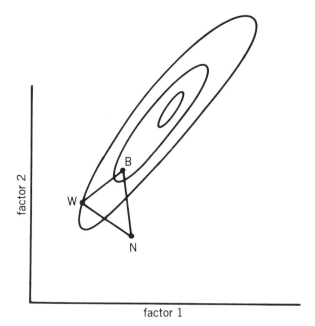

Figure 9.1   Simplex on a response surface.

any one of the three vertices and placing a new vertex anywhere in the factor space. If the new vertex were placed so that it was collinear with the other two vertices, the simplex would have lost one of its dimensions, but this is a special case that we can neglect.

Given that an existent simplex can be changed to another one that is adjacent to the original one, we can make a simplex "walk" about in factor space by a series of replacements. Note that each move of a simplex to create a new, adjacent simplex requires only one evaluation or observation of the response. To start a simplex investigation of a factor space requires $(n+1)$ observations of the response to establish the original simplex. Henceforth, each move requires just one additional observation. The simplex moves about in the factor space quite efficiently, sampling the response values in an orderly way.

To this point, we have not addressed the main question: deciding which direction to move the simplex. We wish to have the simplex move from its current location toward a better region of the factor space. We have the current values of the response where the simplex is now located to work with, and this is all we need. The following paragraphs discuss how the simplex moves through the factor space. The discussion follows that of Deming and Morgan (1973).

Rule 1   The simplex is moved after each observation of the response. Once an initial simplex exists, a move can be made after each additional observation.

Rule 2   A move is made into the adjacent simplex obtained by discarding the vertex of the current simplex corresponding to the least desirable response; it is replaced with its mirror image across the face of the simplex, consisting of the remaining vertices. In Figure 9.2, the original simplex has its vertices labeled B, N, and W. The B vertex corresponds to the best response, N is the next-to-worst response, and W is the worst response. Since W is the worst response, it is discarded, and a new vertex is constructed. The new vertex is found along the line segment connecting W with the centroid of the simplex face of the remaining vertices, here C. The point C, the centroid, is found from the average of the remaining vertices after the worst is discarded. Algebraically, this is

$$C = \frac{P + N}{2}$$

The new vertex is on the line segment connecting W, and C and is placed beyond C at a distance equal to that between W and C. It is labeled R for reflection vertex (Fig. 9.2). Algebraically, this is

$$R = C + (C - W)$$

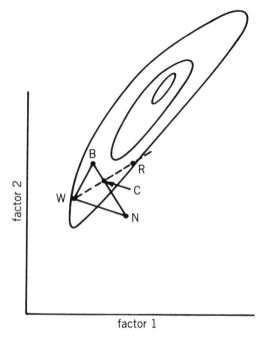

Figure 9.2   Simplex with labeled vertices, including the reflection vertex.

The rejection of the worst vertex forces that simplex to move toward a better part of the response surface.

Figure 9.3 shows a response surface with an initial simplex containing vertices labeled 1, 2, and 3. When vertex 1 is replaced with a reflection vertex, 4, the simplex made its first move in an optimization sequence.

After the application of rule 2, several outcomes are possible. If the response at R is better than that at B or N, rule 2 can be applied again, and another adjacent simplex will be found once again moving the simplex toward the optimum. This loop can continue until an optimum is found. This is what happened in Figure 9.3, where the second simplex (with vertices 2–4) had its worst vertex (2) removed and replaced with the reflection vertex, 5. Then 3 was replaced with 6, and then 4 was replaced with 7. The simplex is making steady progress toward the optimum by moving in a straight line. When the current simplex had vertices 5–7, the direction of movement changed while still using this simple rule.

However, when the response at a reflection vertex is evaluated, and the response at R is the worst response in the new simplex, application of rule 2 would put the simplex back where it immediately came from, and no improvement would be forthcoming, as the simplex would be trapped in an oscillation between two adjacent simplexes. Accordingly, a rule is needed to allow the simplex to break out of this oscillation if necessary.

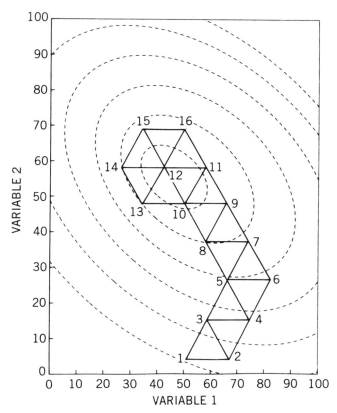

Figure 9.3 Movements of a fixed-size simplex on a response surface. The starting simplex contains vertices 1,2,3, and the optimum point on the response surface is near vertex 12. [Reprinted with permission from *Techniques for the Automated Optimization of HPLC Separations,* John Wiley & Sons, Inc., New York, 1985. Copyright 1985, John Wiley & Sons, Inc.]

Rule 3 If the new vertex found by reflection has the least desirable response in the new simplex, rule 2 does not apply. To continue the series, the next-to-worst response (N) in the new simplex is rejected. This rule allows the simplex to turn as it samples the response surface. This allows the simplex to move as necessary over the response surface seeking the optimum. This rule was invoked in the optimization of Figure 9.3 at the stage where the current simplex was 10, 12, and 13. The worst response of these three vertices is 13, but replacing it with its reflection vertex would return the simplex to the prior one, with vertices 10–12. Therefore, the next-to-worst vertex, 10, was replaced with its reflection vertex, 14.

However, even with the use of rule 3 oscillations can still occur because the simplex can move about a loop in several moves. This behavior is seen in the six simplexes centered on vertex 12. Thus, another rule is needed to eliminate this possibility.

Rule 4 If a particular vertex has been retained in $(k+1)$ simplexes, where k is the dimension of the simplexes, the value of the response at this persistent vertex reevaluated. This is an attempt to ensure that an anomalous observation has not entered into the optimization. If the value of the response at this persistent vertex when reevaluated remains the same, then the optimization is finished. This is what has happened in the optimization of Figure 9.3. No further improvement can be made.

To summarize, Figure 9.3 shows the progress of a simple simplex as it seeks the optimum response on an example surface. It moves largely in a straight line up the response surface and then circles the optimum until it is stopped.

The simple simplex method being described has some limitations that are worth noting. One problem with this approach is that the optimum cannot be found more accurately than the size of the simplex being used. If the simplex is too large, the optimum found will not be well defined. If the simplex is too small, it may take a very large number of moves to find the optimum.

An additional feature that adds utility to the simplex optimization procedure is providing for a way to define regions of the factor space that should not be sampled. For example, if one of the factors being optimized is the pH and if one of the reactants is known to decompose at conditions more acidic than pH = 4.0, then this region of the factor space should be excluded from search. Another example of such a necessity would be an optimization in a chromatographic experiment where the factors being optimized are temperature of the gas-chromatographic oven and carrier-gas flow rate. When both of these factors have small values, then a GC experiment will be very time-consuming. Thus, it might be necessary to restrict search in this region of the factor space. The next rule provides a method for excluding regions of the factor space.

Rule 5 If a new vertex is outside the boundaries of the allowable factor values, a response observation is not made, but instead a very undesirable response value is assigned to this vertex. This will force the simplex back into the allowable region of factor space so that the optimization can continue in the allowed region. This has the effect of putting a "wall" in the way of the simplex so that it cannot go outside the allowed regions of the factor space.

Discussion There are a number of problems with the simple simplex optimization procedure embodied in the rules stated above. Some examples follow. How can one tell when an optimum is reached? What size should the initial simplex be? Will a large initial simplex, which can move through the factor space quickly, find the optimum faster? Or will a small initial simplex, which moves more slowly through the factor space, find the optimum more accurately? How can one be sure that the optimum reached is global? None of these questions have definitive answers. One strategy to use in order to find the optimum region of the factor surface would be to run an optimization with a certain size simplex and then start over in the vicinity of the optimum with a smaller simplex. This will work, but it is not the best way to proceed. Another strategy is to use initial simplexes in different regions of the factor space in a series of optimizations to see if they all converge on the same optimum point in the factor space. In practice, the simple simplex method is not used because of practical difficulties. Some modified methods that skirt some of these problems have been developed, and we discuss them next.

The Variable-Size Simplex Method

Great improvement in the simplex optimization method can be achieved by allowing the simplex to expand and contract as it probes the response surface. Nelder and Mead (1965) developed the variable-size simplex method, which has been called the *modified simplex method* (MSM) in the literature. The expansions or contractions are done during the construction of the adjacent simplex from the existent one in the following way (see Fig. 9.4). Given a simplex labeled B, N, and W as before, construct the reflection vertex R. Now, evaluate the response at R, and perform one of the following operations depending on the value of the response at R. As before, the centroid point is C:

$$C = \frac{P + N}{2}$$

and the reflection vertex is R:

$$R = C + (C - W)$$

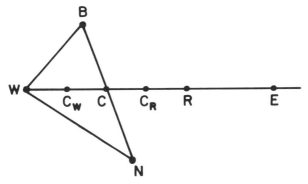

Figure 9.4 The location of the expansion, reflection, and concentration vertices of a simplex.

If the response at R is better than that of B, the simplex is heading in a desirable direction, so try going farther in this same direction. Go to an expansion vertex E:

$$E = C + 2\,(C - W)$$

Usually, the distance from C to E is taken as twice that from C to R. If the response at the expansion vertex E is better than R, save it, and the new simplex will be BNE. If the response at E is not better than R, keep BNR as the new simplex. If the response at R is between that of B and N, the reflection vertex R is kept, and the new simplex is BNR. If the response at R is between that of N and W, a contraction is called for because the simplex is headed in the wrong direction. Depending on how poor the response at R is found to be, the contraction vertex can be placed midway between W and C or midway between R and C. If $N > R > W$, the new vertex is C_R, called the *positive contracted vertex:*

$$C_R = C + 0.5\,(C - W)$$

If $R < W$, the new vertex is C_W:

$$C_W = C - 0.5\,(C - W)$$

Using these rules, the simplex can grow and contract as it moves about the factor space searching for the optimum. It is much less likely to become trapped in local optima than is the fixed-size simplex.

Figure 9.5 shows the progress of the variable-size simplex as it approaches the optimum for the same response surface used in Figure 9.3. The initial simplex is the same as in Figure 9.3. Here, the simplex can move in large steps when it is making good progress toward the optimum. When it nears the optimum, it starts to contract in size and it homes in on the optimum. Along the way there are several vertices (4, 6, and 8) that were tested but were never incorporated into a simplex, so they are isolated.

The final step in simplex optimization is the definition of and use of termination criteria. How does one decide when to stop the optimization procedure? A number of convergence criteria are possible. The one to be used should be chosen by the user knowledgeable of the chemical system involved. Some possible convergence criteria are as follows:

1. Based on the values of the response found. The absolute change or relative change in the response must fall below a threshold. This amounts to looking for level parts of the response surface such as the top of the response surface peak. In addition to the top of peaks, however, there can be other level portions of a surface, for example, inflection points and saddles.

2. Based on the values of the factors. The absolute or relative changes in the values of the factors can be used. This amounts to allowing the simplex size to determine the stopping criterion.
3. Based on small gradients. Numerical evaluations of gradients can be used, but saddle points can fool this type of stopping criterion.
4. Model fitting to several values of the response to try to get a picture of the response surface.

None of these convergence criteria guarantee that the optimization has found a global optimum. None guarantee convergence. In practice, it is usual to follow several criteria simultaneously and to make judgments during the optimization procedure.

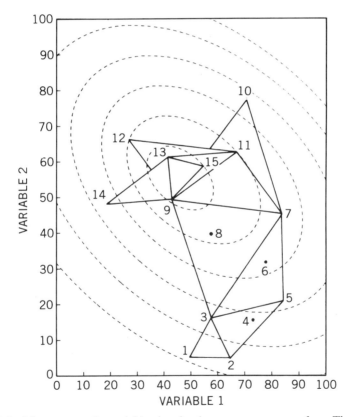

Figure 9.5 Movements of a variable-size simplex on a response surface. The starting simplex contains vertices 1,2,3, and the optimum point on the response surface is within the simplex containing vertices 9,13,15. [Reprinted with permission from *Techniques for the Automated Optimization of HPLC Separations,* John Wiley & Sons, Inc., New York, 1985. Copyright 1985, John Wiley & Sons, Inc.]

9.2 CHEMICAL APPLICATIONS OF SIMPLEX OPTIMIZATION

Morgan et al. (1990) provide an introduction to the simplex optimization method. They illustrate the procedure in great detail as applied to the optimization of a flame atomic absorption spectrometry (AAS) experiment. The experiment involved the determination of chromium in a complex organic matrix by AAS. Two instrumental parameters—the air : fuel ratio and the height of the burner relative to the hollow cathode lamp beam—were optimized. The response optimized was the absorbance measured for standard 5-μg/mL samples. A total of 22 experiments were done and 12 simplexes were generated in identifying the optimum values for the two instrumental variables.

Ernst (1968) was the first to use the simplex method in analytical chemistry in his optimization of NMR magnetic field homogeneity. Gradient and curvature settings of the instrument were varied in a two-dimensional simplex with good results. Chubb et al. (1980) applied simplex optimization to increasing the yield of the Bucherer–Burgs reaction. The reaction is a complex one:

$$R_2C{=}O + NH_3 + COS + HCN \longrightarrow \underset{\substack{}}{\overset{\substack{}}{R_2C}} \begin{array}{c} H \\ \diagup N \diagdown \\ \diagdown C \diagup \\ || \\ S \end{array} \begin{array}{c} O \\ || \\ C \\ | \\ N \\ H \end{array} + H_2O$$

At least eight variables bear on the yield obtained: the initial concentrations of ketone, ammonia, HCN, and carbonyl sulfide plus pH, temperature, and time of the reaction and the solvent. The mixed solvent of ethanol–water was used, with the ratio varied as part of the optimization. Chubb et al. (1980) reported the results of several sets of experiments using cyclohexanone or adamantanone as the starting material. They achieved rapid progress in improving the yields of the reaction using the variable-size simplex optimization procedure. In one series of runs, they improved the yield with cyclohexanone from 49 to 88%. They compared several alternative optimization strategies for their applicability toward organic synthesis.

Morgan and Deming (1975) applied simplex optimization to chromatographic methods development. The goal was the separation of five isomeric octanes, 2,3-dimethylhexane, 3-methylheptane, 2,2-dimethylhexane, 2,2,3,3-tetramethylbutane, and 3,3-dimethylhexane. Packed-column gas chromatography was used, and the two experimental variables under control of the simplex algorithm were column temperature and carrier-gas flow rate. A chromatographic resolution function was used as the response function for the experiments. Dramatic increases in the quality of the chromatograms were seen.

Routh et al. (1977) used simplex optimization in a flame spectroscopy experiment. Calcium emission at 422.7 nm was observed in a nitrous oxide–acetylene flame. The objective was to maximize the net emission signal due to calcium as a function of the following experimental variables: vertical position of burner, horizontal position of burner, fuel flow rate, oxidant flow rate, monochromator setting, monochromator slit width, and photomultiplier high-voltage setting. Rapid convergence to the optimum settings of the variables was observed with several alternative forms of the simplex method, including a variation proposed by the authors called the *supermodified simplex.*

Leary et al. (1982) employed the variable-size simplex method to optimize the instrumental operating conditions for an inductively coupled plasma spectrometer. A multiple-element analysis for Al, Na, Ti, P, and Mn was done with a direct-reading polychromator optically coupled to an argon plasma source. The forward power and the observation height were varied during the simplex optimization. The last simplex found was the best overall for the determination of the five elements, although it did not correspond to the best conditions for any single element.

Berridge (1982) reported the use of the modified sequential simplex algorithm for the unattended optimization of reversed-phase HPLC separations. A chromatographic response function (CRF) was evaluated to represent individual chromatograms with respect to resolution and time of analysis. The CRF contained a number of terms that were derived from the experimentally observed chromatographs. The experiments utilized a microprocessor-controlled liquid chromatograph. Three different sets of experiments were done using two or three variables under investigation. In the first experiment, four 2-substituted pyridines were separated; in the second, a three phenolic antioxidants were separated; in the third, four substituted phenols were separated. The studies showed that completely unattended HPLC optimizations were practical.

Harper et al. (1983) reported using a simplex search to locate an optimum for an ultrasonic extraction of trace elements from atmospheric particulates collected on glass–fiber high-volume sample filters. A method was found that is quantitative for 13 elements and has been designated as an U.S. Environmental Protection Agency (EPA) reference method. The four experimental variables that were optimized were the ratio of hydrochloric to nitric acid, ultrasonic bath temperature, ultrasonic duration, and elapsed time of a diffusion step involved in removing the trapped material from the filter. Studies of the accuracy and precision of recovery of the following elements were reported: As, Ba, Cd, Co, Cr, Cu, Fe, Mn, Mo, Ni, Pb, Sr, Ti, and Zn. The method is being used to collect data for building an archive for all of these elements except Co, Cr, Sr, and Ti.

9.3   NONLINEAR LEAST-SQUARES DATA FITTING BY SIMPLEX

Although the simplex optimization method is well suited to experimental optimization, it can also be used with good results for the fitting of data by nonlinear equations. This is done by setting up the problem so that the response surface that the simplex will investigate is the error function of the data-fitting problem. This is exactly the same error function used in describing curve fitting:

$$Q = \sum_{i=1}^{n} [y_i - f(x_i)]^2$$

The parameters whose values are to be found are contained within the function $f(x)$ in the equation. An advantage of this approach to data fitting is that there is no penalty if the function is nonlinear in the parameters being fit.

Program SIMPLX

Program SIMPLX (see end of chapter) implements the simplex method of optimization for the problem of fitting a set of data with a given function. The program is designed for execution under the immediate supervision of the user. The program begins by asking the user to input the initializing information and the set of data being fit. Then it proceeds to search for the best values of the parameters.

The function subroutine named ERROR evaluates the error function when called by the main routine. In this implementation of the simplex procedure, the better response is the smaller value since an error is being minimized. The statement in which the variable YCALC is computed is where the program must be changed in order to use a different function from that used in this example. The function being used here as an example is

$$A = A_\infty (1 - e^{-kt})$$

where $A\infty$, and k are to be found using simplex optimization.

Program SIMPLX is set up as an accommodation between flexibility and simplicity. Several of the program parameters have been defined within the program, although they could be supplied at execution time by the user if desired. For example, MAXCNT is the number of error function evaluations that are allowed before the program will abort execution on the assumption that no further progress is possible. ERRMIN is given a value of 1.0×10^{-3} inside the program, but this, too, could be supplied by the user at execution time if desired. The program has several extra WRITE statements that will only be executed if the user answers "yes" to the debug output question.

The stopping criterion implemented in program SIMPLX is the following. When the fractional change in response surface values between the best vertex and the worst vertex in the current simplex falls below the value in ERRMIN, the routine terminates.

The set of data being fit in the example execution of the program is taken from a literature paper by Deming and Morgan (1973). The values of the data set are as follows:

t	A	t	A
1.5	0.110	9.0	0.325
1.5	0.109	12.0	0.326
3.0	0.169	12.0	0.330
3.0	0.172	15.0	0.362
4.5	0.210	15.0	0.383
4.5	0.210	18.0	0.381
6.0	0.251	18.0	0.372
6.0	0.255	24.0	0.422
9.0	0.331	24.0	0.411

Program SIMPLX was executed twice with this same set of data but with different starting values for the two parameters in the equation. In the first run, the starting values were provided as 0.5 and 1.0. The value of the error function for this set of parameter values was 0.8194. As can be seen from the output of this run, the error value rapidly decreased as the simplex procedure homed in on the best values. A great deal of the output has been deleted in the interest of saving space. The final results obtained were $A_\infty = 0.404$, $k = 0.170$, and an error value of 0.0036, which are all identical with the values reported in the original paper. In the second run, the starting values were provided as 1.0 and 0.5. The starting value for the error function was 7.009 in this case. Again, the error rapidly decreased as the simplex homed in on the best values. The final results were identical with the first run in all respects.

```
      PROGRAM SIMPLX
C....
C.... SIMPLEX MINIMIZATION OF A FUNCTION
C....
      DIMENSION C(10),E(10),P(10,10),R(10),X(10)
      DIMENSION DATA(100,10)
C.... ENTER INITIAL INFORMATION
      MAXCNT=500
      ERRMIN=1.0E-03
      NOUT=6
      NINP=5
      NSIM=1
      WRITE (NOUT,*) ' THIS IS PROGRAM SIMPLX'
      WRITE (NOUT,*) ' ENTER NUMBER OF OBSERVATIONS'
      READ (NINP,*) NOBS
      WRITE (NOUT,*) ' ENTER NUMBER OF VARIABLES (X+Y)'
      READ (NINP,*) NV
      WRITE (NOUT,*) ' ENTER X1,X2,...,XP AND Y'
      DO 6 I=1,NOBS
    6 READ (NINP,*) (DATA(I,J),J=1,NV)
      IF (NOBS.GT.NV) GO TO 11
      WRITE (NOUT,9)
```

```
    9 FORMAT (' NO. OF OBS. MUST BE GT NO. OF VARS.')
      STOP
   11 WRITE (NOUT,*) ' ENTER NUMBER OF PARAMETERS'
      READ (NINP,*) NP
      NP1=NP+1
      WRITE (NOUT,*) ' ENTER INITIAL ESTIMATES OF PARAMETERS'
      READ (NINP,*) (X(I),I=1,NP)
      E(1)=ERROR(X,DATA,NV,NOBS,KOUNT)
      WRITE (NOUT,15) E(1)
   15 FORMAT (' STARTING ERROR FUNCTION VALUE',G12.4)
      WRITE (NOUT,*) ' WANT DEBUG LEVEL OUTPUT (Y OR N)'
      READ (NINP,18) ANS
   18 FORMAT (A1)
      IDB=0
      IF (ANS.EQ.'Y') IDB=1
C.... INITIALIZE THE SIMPLEX
      KOUNT=0
      DO 22 J=1,NP
   22 P(1,J)=X(J)
      DO 28 I=2,NP1
         DO 26 J=1,NP
   26    P(I,J)=X(J)
         P(I,I-1)=1.1*X(I-1)
         IF (ABS(X(I-1)).LT.1.0E-12) P(I,I-1)=0.0001
   28 CONTINUE
C.... FIND PLOW AND PHIGH/ BEST=PLOS/ WORST=PHIGH
   31 ILO=1
      IHI=1
      DO 34 I=1,NP1
         DO 32 J=1,NP
   32    X(J)=P(I,J)
         E(I)=ERROR(X,DATA,NV,NOBS,KOUNT)
         IF (E(I).LT.E(ILO)) ILO=I
         IF (E(I).GT.E(IHI)) IHI=I
   34 CONTINUE
      WRITE (NOUT,*) ' INITIAL SIMPLEX'
      DO 40 K=1,NP1
      WRITE (NOUT,39) K,E(K),(P(K,J),J=1,NP)
   39 FORMAT (3X,' VERTEX',I2,' ERROR AND PARAMETERS:',5F8.3)
   40 CONTINUE
C.... FIND PNHI THE NEXT HIGHEST  NEXT=PNHI
   41 NHI=ILO
      DO 43 I=1,NP1
         IF (E(I).GE.E(NHI).AND.I.NE.IHI) NHI=I
   43 CONTINUE
C.... COMPUTE THE CENTROID
      DO 46 J=1,NP
         C(J)=-P(IHI,J)
         DO 44 I=1,NP1
            C(J)=C(J)+P(I,J)
   44    CONTINUE
         C(J)=C(J)/NP
   46 CONTINUE
   51 CONTINUE
C.... PRINT CURRENT BEST VERTEX
      WRITE (NOUT,53) KOUNT,NSIM
   53 FORMAT (' AFTER',I3,' ERROR EVALUATIONS AND',I3,' SIMPLEXES')
      WRITE (NOUT,54) (P(ILO,J),J=1,NP)
   54 FORMAT ('    PARAMETER ESTIMATES:',5G12.4)
      WRITE (NOUT,55) E(ILO)
   55 FORMAT ('    ERROR FUNCTION:',G12.4)
C.... STOPPING CRITERION
      IF (KOUNT.GT.MAXCNT) STOP
      IF (ABS(E(ILO)-E(IHI))/E(ILO).LT.ERRMIN) GO TO 56
      GO TO 61
```

```
    56 WRITE (NOUT,*) ' ==> ERROR CRITERION SATISFIED'
       WRITE (NOUT,54) (P(ILO,J),J=1,NP)
       STOP
C.... REFLECTION
    61 DO 62 J=1,NP
          R(J)=1.9985*C(J)-0.9985*P(IHI,J)
    62 CONTINUE
       ER=ERROR(R,DATA,NV,NOBS,KOUNT)
       IF (IDB.GT.0) WRITE (NOUT,65) ER,(R(J),J=1,NP)
    65 FORMAT (' REFLECTION VERTEX',3F10.5)
C.... REFLECT AGAIN IF SUCCESSFUL
       IF (ER.LT.E(ILO)) GO TO 91
       IF (ER.GE.E(IHI)) GO TO 122
C.... REPLACE WORST VERTEX WITH NEW ONE
    79 DO 80 J=1,NP
          P(IHI,J)=R(J)
    80 CONTINUE
       NSIM=NSIM+1
       E(IHI)=ER
       IF (ER.GT.E(NHI)) GO TO 51
       IHI=NHI
       GO TO 41
C.... EXPAND THE SIMPLEX
    91 ILO=IHI
       IHI=NHI
       DO 93 J=1,NP
          X(J)=1.95*R(J)-0.95*C(J)
    93 CONTINUE
       EX=ERROR(X,DATA,NV,NOBS,KOUNT)
       IF (EX.LT.ER) GO TO 104
C.... R BETTER THAN X
       DO 99 J=1,NP
          P(ILO,J)=R(J)
    99 CONTINUE
       NSIM=NSIM+1
       E(ILO)=ER
       GO TO 110
C.... X IS BETTER THAN R
   104 DO 105 J=1,NP
          P(ILO,J)=X(J)
   105 CONTINUE
       IF (IDB.GT.0) WRITE (NOUT,106) EX,(X(J),J=1,NP)
   106 FORMAT (' EXPANSION VERTEX',3F10.5)
       NSIM=NSIM+1
       E(ILO)=EX
   110 CONTINUE
       GO TO 41
C.... CONTRACT THE SIMPLEX
   122 DO 123 J=1,NP
          R(J)=0.5015*C(J)+0.4985*P(IHI,J)
   123 CONTINUE
       ER=ERROR(R,DATA,NV,NOBS,KOUNT)
       IF (IDB.GT.0) WRITE (NOUT,124) ER,(R(J),J=1,NP)
   124 FORMAT (' CONTRACTION VERTEX',3F10.5)
       IF (ER.LT.E(ILO)) GO TO 91
       IF (ER.LT.E(IHI)) GO TO 79
C.... SCALE
       WRITE (NOUT,135)
   135 FORMAT (' ENTER SCALE (<0 EXPANDS, >0 SHRINKS, 0=STOP):')
       READ (NINP,*) SCAL
       IF (SCAL.EQ.0.0) GO TO 999
   137 DO 138 I=1,NP1
          DO 138 J=1,NP
             P(I,J)=P(I,J)+SCAL*(P(ILO,J)-P(I,J))
```

```
    138 CONTINUE
        GO TO 31
    999 STOP
        END
C-------------------------------------------------------------
        FUNCTION ERROR (X,DATA,NV,NOBS,KOUNT)
C.... COMPUTES THE ERROR FUNCTION FOR THE DATA SET
C....    SMALLER VALUE IS BETTER
        DIMENSION X(10),DATA(100,10)
        ERROR=0.0
        DO 10 I=1,NOBS
          YOBS=DATA(I,NV)
C.... CHANGE THE NEXT STATEMENT TO CHANGE THE FUNCTION BEING FIT
          YCALC=X(1)*(1.0-EXP(-X(2)*DATA(I,1)))
          RESI=YOBS-YCALC
          ERROR=ERROR+RESI*RESI
     10 CONTINUE
        KOUNT=KOUNT+1
        RETURN
        END
```

```
THIS IS PROGRAM SIMPLX

ENTER NUMBER OF OBSERVATIONS
18
ENTER NUMBER OF VARIABLES (X+Y)
2
ENTER X1,X2,...,XP AND Y
1.5    0.110
1.5    0.109
3.0    0.169
3.0    0.172
4.5    0.210
4.5    0.210
6.0    0.251
6.0    0.255
9.0    0.331
9.0    0.325
12.0   0.326
12.0   0.330
15.0   0.362
15.0   0.383
18.0   0.381
18.0   0.372
24.0   0.422
24.0   0.411
ENTER NUMBER OF PARAMETERS
2
ENTER INITIAL ESTIMATES OF PARAMETERS
0.5  1.0
STARTING ERROR FUNCTION VALUE   0.8194
WANT DEBUG LEVEL OUTPUT? (Y OR N)
N

INITIAL SIMPLEX
    VERTEX 1 ERROR AND PARAMETERS:    0.819   0.500   1.000
    VERTEX 2 ERROR AND PARAMETERS:    1.204   0.550   1.000
    VERTEX 3 ERROR AND PARAMETERS:    0.848   0.500   1.100
AFTER   3 ERROR EVALUATIONS AND   1 SIMPLEXES
    PARAMETER ESTIMATES:  0.5000      1.000
    ERROR FUNCTION:  0.8194
```

10

CHEMICAL STRUCTURE
INFORMATION HANDLING

10.1 INTRODUCTION

The least common denominator of communication between chemists regarding molecular structures of compounds is the standard two-dimensional structural diagram. Blackboard sketching is an art form in chemistry. This type of sketched structure provides visual communication of concepts that would be extremely cumbersome to describe orally. So much is this "language" embedded within chemistry that if you draw a hexagon for a chemist with no explanation at all, and ask what it is, you are likely to get the reply "cyclohexane." Imagine an organic chemistry text without structural diagrams, and the importance of this "language" is even more apparent. These two-dimensional structural diagrams are an abstraction, and they have very little relation to the fundamental nature of the species they represent. They are a shorthand description, not an explanation.

However, standard two-dimensional structural diagrams for chemical compounds are inadequate for many operations of modern chemistry. Examples are abundant, and journals commonly print structures in different ways to demonstrate the three-dimensional aspects of structure. Especially for the input, storage, manipulation, and display of chemical structures within computer-based systems, some more powerful representation must be chosen. It must be compatible with the chemist user's knowledge and background and also with the demands of computers. A number of different representations have been developed over the past 20 years. The selection of a representation to be used in a specific set of circumstances depends on many factors, including the following: the size of the files to be handled (hundreds or tens of thousands of compounds), the functions to be performed on the files (sort and print lists, prepare indexes, substructure searching), the available hardware (micro-, mini-, or mainframe computer), the available software, the degree of automation desired, and the knowledgeability of the personnel using and maintaining the system. The method of structural representation chosen is critical to much of the remainder of the system in a myriad of ways.

Two important characteristics of a structural representation are (1) uniqueness, that is, only one representation can be derived from a compound; and (2) unambiguousness, a representation applies to only one compound. These two characteristics are complementary, since one applies to encoding and the other to decoding of a structural representation. Additionally, a representation must be complete and therefore describe the entire structural diagram. For practical reasons, it is also desirable for a representation to be concise, using as little storage space as possible.

One scheme for classifying structural representations of molecules is given in Figure 10.1. Fragment codes represent structures by noting the presence of predetermined selected portions of them, such as functional groups and ring systems. They are ambiguous and are important primarily as the historical antecedents of the present systems. The unambiguous representations of several classes will be dealt with here in detail. Topological representations record only the topology or connectivity of the molecule. That is, they record the atom types, bond types, and interconnections among the atoms. Extensions

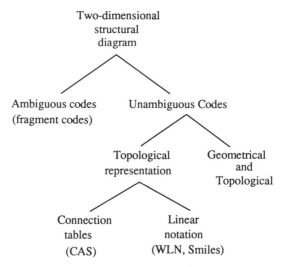

Figure 10.1 Chemical structure representations. [Reprinted with permission from *Computer Handling of Chemical Structure Information*, Macdonald, London, 1971. Copyright 1971, Macdonald.]

can also record absolute configurations or stereochemistry about asymmetric centers. There are two main classes of topological representations: linear notations and connection tables.

Linear notations condense three-dimensional molecular structures into a unique, unambiguous one-dimensional string of symbols. The notations are compact (an advantage when stored) because most of the bonds are not recorded explicitly, but implicitly. The notation, the sequence of symbols, for a given structure is produced by following a set of rules. Linear notations were originally designed for manual use, but they were readily adaptable for use with punched-card sorting equipment (late 1940s) and computers. While a number of linear notations have been developed (Wiswesser, IUPAC, Hayward), the Wiswesser line notation (WLN) has been the most widely adopted. However, the line notations have fallen into disuse as graphically oriented structure entry systems have become more widely available.

10.2 CONNECTION TABLE REPRESENTATION

The topology of molecular structures can be explicitly represented in the form of connection tables. To generate the connection table for a compound, the nonhydrogen atoms are numbered arbitrarily, and a table is constructed that cites all the interconnections. Table 10.1 shows the simple connection table for the seven-atom molecule, 3-aminobutanoic acid:

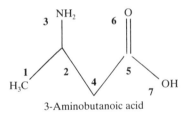

3-Aminobutanoic acid

TABLE 10.1   Connection Table for 3-Aminobutanoic Acid

				Connections			
Atom Number	Atom Type	Atom Number	Atom Type	Atom Number	Atom Type	Atom Number	Atom Type
1	C	2	1	—	—	—	—
2	C	1	1	3	1	4	1
3	N	2	1	—	—	—	—
4	C	2	1	5	1	—	—
5	C	4	1	6	2	7	1
6	O	5	2	—	—	—	—
7	O	5	1	—	—	—	—

In the usual case, the hydrogen atoms are not cited explicitly. In the simplest connection table, each bond has been cited twice, which constitutes unnecessary and undesirable redundancy. This redundancy can be eliminated, without loss of information, by adopting the following two conventions:

1. In numbering the structure, once an atom has been numbered, all unnumbered atoms connected to it are numbered serially.
2. In the connection table, cite connections only to lower numbered atoms.

Applying these rules to the same seven-atom molecule yields the compact connection table of Table 10.2. The numbering sequence used in Table 10.1 followed the numbering convention, so renumbering was not necessary in this example.

For cyclic structures, one extra line must be added to the compact connection table for each ring. This ring-closure entry specifies the one extra bond found because of the presence of the ring.

Aromatic rings can be handled in several ways. Either the bond types can be entered as alternating single and double bonds, or a special bond type can be adopted for aromatic bonds only. Some connection table conventions follow the first definition and some follow the second.

Evidently, many types of enumeration are possible for any given structure. For a molecule of n atoms (not counting hydrogen atoms), there are $n!$ possible numbering schemes. Each of these numbering schemes will yield a different connection table. Thus, the connection tables generated by the procedure described above are not unique. However, they do completely specify the topology of the molecular structure, and therefore they are unambiguous.

Charges, abnormal valences, and so on, are stored as adjunct information in addition to the connection table.

The information stored in connection tables can be stored in an alternative mode utilizing binary matrices. Thus, the same information stored in the connection table shown in Table 10.2 could be stored as shown in Table 10.3.

As will be shown in a later section, structures represented by connectivity

TABLE 10.2   Compact Connection Table for 3-Aminobutanoic Acid

Atom Number	Atom Type	Connections	
		Atom Number	Atom Type
1	C	—	—
2	C	1	1
3	N	2	1
4	C	2	1
5	C	4	1
6	O	5	2
7	O	5	1

11

MATHEMATICAL GRAPH THEORY

11.1 INTRODUCTION

Graph theory is the study of the nature and properties of topological graphs. This mathematical discipline is closely related to both topology and combinatorics, and it has a multitude of applications to chemistry. The origins of graph theory date back over 200 years to the work of individuals from several different fields. Euler (1707–1783) is generally considered to be the originator of graph theory with his publication of the solution of the Königsberg Bridge problem. Kirchhoff's work with electric circuits, and Cayley's work leading to his enumeration of organic chemical isomers are recognized as independent discoveries of graph theory (Harary 1969).

Graph theory has found application throughout science and engineering and beyond. Within chemistry, graph theory has been applied to problems from a wide range of research areas, including synthetic chemistry, polymer chemistry, quantum chemistry, thermochemistry, chemical kinetics, statistical mechanics, phase equilibria, spectroscopic analysis, Hückel theory, and chemical information storage and retrieval.

In the sense used here, the term *graph* must not be confused with the familiar Cartesian coordinate plots (viz., y vs. x) of experimental data. In graph theory, a graph is an abstract concept applied to a collection of points (also called nodes or vertices) joined by lines (also called edges). Edges connect pairs of vertices together. Thus, graphs can be shown pictorially, and some examples of graphs are shown in Figure 11.1. More formally, a graph is a pair of sets: (1) a set of elements and (2) a set of pairs of these elements.

Some of the fundamental definitions of graph theory follow. A graph is a

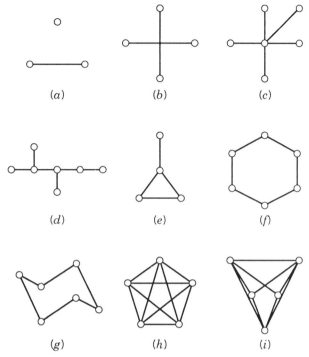

Figure 11.1 Examples of graphs. Graphs *a* and *b* are disconnected, but all the other graphs are connected. Graphs *c* and *d* are trees. Graphs *e*–*i* are cyclic. Graphs *f* and *g* are isomorphic, as are graphs *h* and *i*. Examples of a regular graph (all of whose points have the same degree) are *b* and *f*–*i*. Graphs *h* and *i* are complete graphs (every pair of points is adjacent).

finite collection of points (also called *vertices* or *nodes*) and a finite collection of lines (also called *edges*). The edges connect pairs of vertices together. Two vertices connected by an edge are *adjacent*. A *walk* of a graph is an alternating sequence of points and lines that begins and ends with points. The walk becomes a *path* if all the points traversed are distinct. The walk is a *cycle* if it is closed and if at least three distinct points are traversed. The *length* of a walk is the number of lines in it. The *distance* between two points in a graph is the length of the shortest path joining them. The *degree* of a point is the number of lines connected to it. A graph is *regular* if every point in it has the same degree. A graph is *connected* if every pair of points is joined by a path. A graph is *complete* if every point is connected to every other point. A graph is *acyclic* if it has no cycles. A *tree* is a connected, acyclic graph. A graph is *labeled* when its nodes are distinguishable from one another by names. Two graphs are *isomorphic* if there exists a one-to-one correspondence between their points and lines. Isomorphic graphs can present quite different appearances when represented pictorially. Many of these concepts are illustrated in the graphs of Figure 11.1.

An obvious use for graph theory is for depicting and manipulating molecular structures. The points of the graph are the atoms of the molecule, and the edges of the graph are the bonds of the molecule. The degree of an atom is the number of bonds connected to it, the valence.

Figure 11.2 shows pictorial representations of all 23 trees that have eight nodes. Five of these tree graphs are chemically implausible (as carbon-backbone compounds) because they contain nodes of degree greater than four (>4). Thus, if four is the maximum degree allowed, there are exactly 18 possible tree structures with eight nodes. If the nodes are allowed to be carbon atoms only and the edges are allowed to be single bonds only, then this statement can be revised to reflect the fact that there are exactly 18 acyclic alkanes with eight carbons.

The structure of a graph can be represented by an adjacency matrix, $\mathbf{A} = a_{ij}$, where $a_{ij} = 1$ if nodes i and j are adjacent and 0 otherwise. Figure 11.3

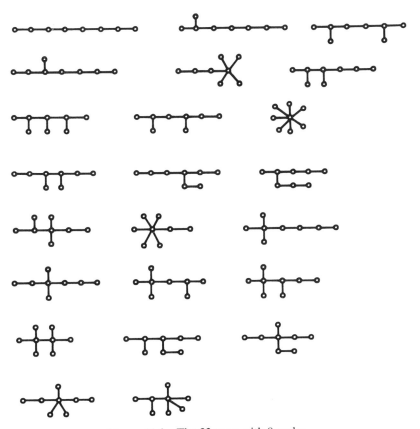

Figure 11.2 The 23 trees with 8 nodes.

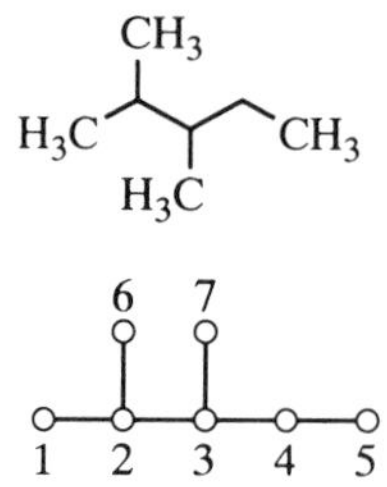

$$A = \begin{bmatrix} 0 & 1 & 0 & 0 & 0 & 0 & 0 \\ 1 & 0 & 1 & 0 & 0 & 1 & 0 \\ 0 & 1 & 0 & 1 & 0 & 0 & 1 \\ 0 & 0 & 1 & 0 & 1 & 0 & 0 \\ 0 & 0 & 0 & 1 & 0 & 0 & 0 \\ 0 & 1 & 0 & 0 & 0 & 0 & 0 \\ 0 & 0 & 1 & 0 & 0 & 0 & 0 \end{bmatrix}$$

$$D = \begin{bmatrix} 0 & 1 & 2 & 3 & 4 & 2 & 3 \\ 1 & 0 & 1 & 2 & 3 & 1 & 2 \\ 2 & 1 & 0 & 1 & 2 & 2 & 1 \\ 3 & 2 & 1 & 0 & 1 & 3 & 2 \\ 4 & 3 & 2 & 1 & 0 & 4 & 3 \\ 2 & 1 & 2 & 3 & 4 & 0 & 3 \\ 3 & 2 & 1 & 2 & 3 & 3 & 0 \end{bmatrix}$$

Figure 11.3 2,3-Dimethylpentane represented as a structural graph, an abstract graph, adjacency matrix, and distance matrix.

shows 2,3-dimethylpentane represented pictorially as a graph and also shows the corresponding adjacency matrix **A**. This is the same as the binary atom connectivity matrix introduced in Chapter 10. Thus, a direct link exists between the representation of chemical structures as connection tables and mathematical graph theory. Theorems developed in the abstract domain of graph theory can be applied directly to chemical structures in many cases. For example, the row sums of the adjacency matrix are the connectivities of the atoms. The number of paths of length n within a molecule can be found by calculating A^n. The value for the i,j entry in the resulting matrix is the number of paths of length n from atom i to atom j.

Once the connection between graph theory and chemical structure representation has been established, the theorems of graph theory can be exploited for chemical purposes. This has led to important advances in a number of areas of chemistry, including the following:

1. Registration of chemical compounds.
2. Isomer enumeration.
3. Ring locating.

4. Substructure searching.

5. Quantification of structural similarity.

Many reviews of the applications of graph theory to chemical problems have appeared in the literature (e.g., Balaban 1976, 1985, Rouvray 1986).

11.2 REGISTRATION OF CHEMICAL COMPOUNDS

The ability to store individual chemical compounds in a computer-compatible form using a linear notation or connection-table representation leads to the ability to generate files of compounds. In many industrial companies, such compound files contain hundreds of thousands of structures, often with associated data such as spectra, physical properties, or test results. In the process of building such files of chemical structures, when each compound to be added to the file is being processed, it is necessary to answer the question, "Is this compound already in the chemical structure file?" In other words, is the compound being handled really new, or is it already present in the file. If the answer is "yes," then the compound need not be added. If the answer is "no," a new entry is necessary. The process of answering this question is called *registration*. The ability to perform the registration function accurately and efficiently is crucial in systems such as the Chemical Abstracts Service registry system. In this system, information relating to a compound is stored under the compound's identifying name.

If the compound is represented by a linear notation, answering this registration question is equivalent to performing a search through the file because the notation is unique. If the compound were already in the file, it would have been represented by precisely the same linear notation when originally entered. Therefore, searching the files for the query compound amounts to looking for the same string of symbols that represent the query compound in the already existent file.

However, as we have seen in Chapter 10, a simple connection-table representation of a chemical structure is not unique because the architecture of the connection table depends on the numbering of the non-H atoms in the structure. To make it unique, which greatly simplifies the registration problem, algorithms based on graph theory have been developed. The most important one is known as the *Morgan algorithm* (Morgan 1965), after its developer.

A given chemical compound can be represented by many equivalent connection tables, depending on how the numbers were assigned to the atoms. Generating a unique connection table, a canonical connection table, is done by generating a unique, invariant numbering for the structure based on its own inherent structure. If any arbitrary numbering sequence is allowed, there are $n!$ (n factorial) numbering schemes for a molecule with n non-H atoms.

This presents a gargantuan number of potential numbering sequences for molecules of even moderate size. The potential number of sequences can be reduced dramatically by following the numbering rules presented in Section 10.2. That is, once an atom has been numbered, then all unnumbered atoms connected to it are numbered serially. While the number of sequences that can arise from this scheme is not easily calculated, it is still extremely large. There must be a method for choosing just one numbering sequence in a repeatable way. The Morgan algorithm utilizes the structure of the compound being treated to resolve this problem. It focuses on the extended connectivities of the atoms (the number of non-H atoms connected to them) and the atom types to sort the atoms. Expressed in words, the algorithm seeks the most deeply imbedded atom within the structure to identify the atom to be given the number 1. Then the procedure follows the serial numbering rule, resolving ties by the extended connectivities of the atoms involved, until the structure is completely numbered. In algorithm form, the procedure is as follows (O'Korn 1977):

1. Calculate stage 1 connectivity values for each atom by inspecting the structure.

2. Calculate stage 2 connectivity values for each atom by summing the stage 1 connectivity values for the attached atoms.

3. Calculate stage $i + 1$ connectivity values for each atom by summing the stage i connectivity values for the attached atoms.

4. At each stage, calculate the number of distinct connectivity values that appear in the structure, k.

5. As long as the number of distinct connectivity values continues to increase, keep repeating steps 3 and 4.

6. When a stage is found where the number of distinct connectivity values stays equal or decreases, save the connectivity values from the previous stage.

7. The atom with the highest connectivity value is given the number 1.

8. Note all other atoms with the same connectivity value for later use.

9. Serially number all atoms connected to atom 1 as number 2, 3, and so on, based on decreasing connectivity values. If an arbitrary choice must made, note the pairs of atoms involved in the arbitrary choice for later use.

10. The unnumbered atoms connected to atom 2 are serially numbered based on decreasing connectivity values. As before, note the pairs of atoms involved in any arbitrary choices.

11. Follow this procedure until all the atoms have numbers.

12. Generate the compact connection table based on this numbering sequence. If no arbitrary choices were made during the numbering, this connection table is the canonical one, and the procedure is terminated.

13. If there were arbitrary choices made, back up to the highest numbered atom for which an arbitrary choice was made.

14. Select the other atom from the pair involved in the arbitrary choice, and renumber the atoms of the structure from that atom to the last atom.

15. Generate the compact connection table based on this numbering sequence.

16. Compare the newly generated connection table to the previously generated one.

17. If the new connection table is better than the previous one, replace the retained one with the new one.

18. Go back to step 13.

In step 17 the concept of "better" was introduced. To generate canonical numbering, it is not necessary for the definition of better to be sophisticated, only invariant. In the original formulation, alphabetic ordering was used. This is equivalent to using atom types to decide precedence.

Figure 11.4 shows the application of the Morgan algorithm to the nine-atom molecule, 2-isopropyl-4-methyltetrahydrofuran:

The initial set of connectivities contains the three values (1,2,3) and $k = 3$. By stage 3, k has increased to 6, and it does not increase when an additional step is taken. Therefore, the stage 3 connectivity values are taken as the final ones for use during numbering. During numbering, an arbitrary choice must be made when assigning the number 3. Therefore, two different sequences have been generated, and two corresponding compact connection tables can be generated. The arbitrary choice in assigning the number 5 is unimportant because atoms 5 and 6 are identical through symmetry. The two compact connection tables are compared to determine which is better, and the one corresponding to the oxygen atom numbered 4 is chosen because it is desirable to have heteroatoms appear late in the connection table. The actual canonical "name" of this compound to be used in the registration process is then formed from this preferred compact connection table. The string of symbols formed by sweeping out the connection table is unique and can be used directly in searching files for a match.

The important point about this scheme for generating the canonical numbering for any molecule is that the same molecule will always generate exactly

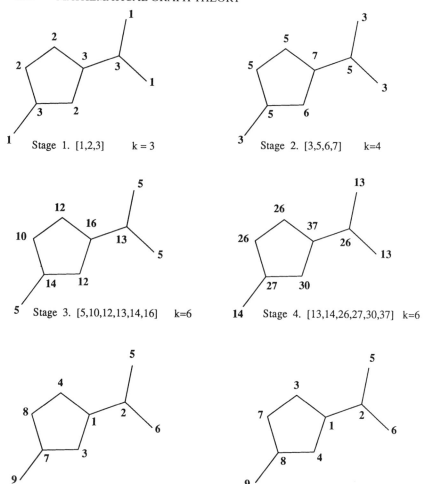

Figure 11.4 Application of the Morgan algorithm to 2-isopropyl-4-methyltetra-hydrofuran.

the same connection table. Then, using the connection table, it is straightforward to compare the connection table to others stored in the structure files to determine whether an identical connection table is already present.

A number of extensions of this basic algorithm have been developed to deal with additional complexities of molecular structures. For example, the stereochemistry of structures can be handled by using extensions described by Wipke and Dyott (1974).

Atom Number	Atom Type	Connections	
		Atom Number	Bond Type
1	C	--	--
2	C	1	1
3	C	1	1
4	O	1	1
5	C	2	1
6	C	2	1
7	C	3	1
8	C	4	1
9	C	7	1

(*b*)

Figure 11.4.   (*Continued*)

11.3   ENUMERATION OF ISOMERS

The first work on isomer enumeration using graph theory was by Cayley in 1874, and it dealt with tree structures (Cayley, 1874). Techniques involving recursion formulas were first developed by Henze and Blair (1931–1932). This approach involves using the isomer count for a given number of homologous series containing n carbon atoms and then computing the isomer count for the succeeding member containing $n + 1$ carbon atoms. Many authors have applied this recursive approach to a variety of chemical systems (Rouvray 1974). These methods are particularly well suited to computer implementation.

The single most important advance in isomer enumeration was Polya's enumeration theorem (Polya 1937), which exploited group theory. It is presented and discussed in many secondary sources (e.g., Harary 1969, Harary et al. 1976).

The field of isomer enumeration continues to attract the attention of mathematical chemists. Lederberg and colleagues developed a general procedure for enumerating the acyclic isomers corresponding to a given elemental composition (Lederberg et al. 1969), and the algorithm, named DENDRAL and later renamed CONGEN, has been improved several times. The research group that developed these approaches has reported an algorithm capable of generating not only constitutional isomers but stereoisomers as well (Nourse et al. 1980).

11.4   ATOMIC AND MOLECULAR PATH COUNTS

The use of graph theoretic properties of chemicals represented as graphs for chemical purposes has attracted a great deal of attention. Another example is that of atomic and molecular paths. In the context of graph theory, a "path" is defined as a sequence of nodes in a graph that are connected by edges, where no node appears more than once in the list. The length of a path is the number of edges in the path. Since organic molecules can be represented as graphs, their paths can be investigated as a chemical property. The enumeration of the paths in a molecule becomes progressively more difficult as the molecule increases in size, in connectivity, or both.

To illustrate the idea of paths within molecules, consider the example molecule shown in Figure 11.5. The molecule is drawn in the usual two-dimensional structural diagram, then with the hydrogen atoms suppressed and the atoms

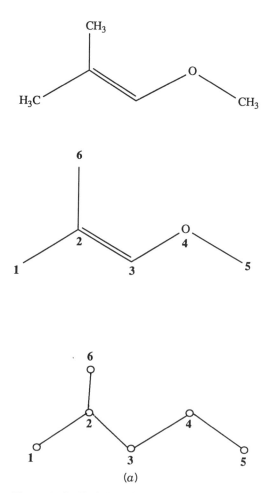

(a)

Figure 11.5   Path lengths for example molecule.

numbered, and finally as a numbered graph with nodes and edges. Focus first on atom 1. There is only one path of length 1, namely 1–2. There are two paths of length 2, namely 1–2–3 and 1–2–6. There is one path of length 3 and one path of length 4. For atom 1 of the example molecule, the atom path code is 1,2,1,1. For the other atoms in the structure, the atom path codes are different, according to the degree to which they participate in paths of different lengths.

In acyclic structures, there is a unique path between any pair of atoms, so the number of paths of a specified length gives the number of neighbors at that distance. The number of neighbors a given distance away from the atom of interest is the basis for several schemes for ^{13}C NMR shift predictions by additivity methods (e.g., Grant and Paul 1964, Lindeman and Adams 1971).

Figure 11.6 shows the atom path codes for all five isomers of hexane. Each of these isomers has a total of 15 unique paths, but they are distributed differently among the possible lengths. Some regularities are evident in the codes for these structures. For example, the length of the longest path for each atom in a structure is the number of bonds that must be traversed to go from that atom to the most distant terminal atom. Only a completely extended alkane of n atoms with no internal branching can have a path of length $n - 1$. There is a general relationship between the "extent" of a molecule and the length of the paths found within it.

Another representation based on paths, the molecular path code, is obtained by summing the numbers of atomic path codes by lengths. Each path appears twice, once for each end atom, so the results obtained in the summing operation must be divided by 2. Generally, different molecules have different

Atom Number	Path Length				
	1	2	3	4	5
1	1	2	1	1	0
2	3	1	1	0	0
3	2	3	0	0	0
4	2	1	2	0	0
5	1	1	1	2	0
6	1	2	1	1	0
Molecular Path Count	5	5	3	2	0

(*b*)

Figure 11.5   (*Continued*)

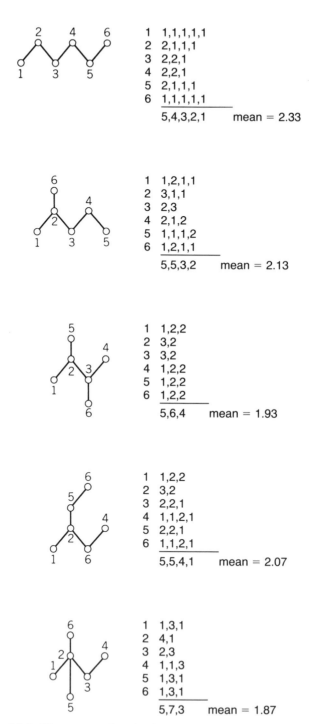

1 1,1,1,1,1
2 2,1,1,1
3 2,2,1
4 2,2,1
5 2,1,1,1
6 1,1,1,1,1
_____
5,4,3,2,1   mean = 2.33

1 1,2,1,1
2 3,1,1
3 2,3
4 2,1,2
5 1,1,1,2
6 1,2,1,1
_____
5,5,3,2   mean = 2.13

1 1,2,2
2 3,2
3 3,2
4 1,2,2
5 1,2,2
6 1,2,2
_____
5,6,4   mean = 1.93

1 1,2,2
2 3,2
3 2,2,1
4 1,1,2,1
5 2,2,1
6 1,1,2,1
_____
5,5,4,1   mean = 2.07

1 1,3,1
2 4,1
3 2,3
4 1,1,3
5 1,3,1
6 1,3,1
_____
5,7,3   mean = 1.87

Figure 11.6 The atom path codes for all five isomers of hexane.

molecular path codes. The molecular path code for each of the five hexane isomers is shown in Figure 11.6. The average path length for each isomer is also shown. The most extended isomer, *n*-hexane, has the largest average value of 2.33; and the most highly branched isomer, 2,2-dimethylbutane, has the smallest average path length of 1.87. The hexane isomers of intermediate branching have intermediate average path lengths, as shown in Figure 11.6.

Molecular path codes provide a method for studying the degree of similarity between molecules quantitatively (Randić and Wilkins 1979). Randić and Wilkins computed the structural similarities among a set of 29 monocyclic monoterpenes by evaluating the following dissimilarity measure:

$$D_{ab}^2 = \sum_i (a_i - b_i)^2$$

where *a* and *b* refer to the molecular path codes for two different structures and the summation is carried out over the index *i*, which indicates the lengths of the paths. Here $D = 0$ for molecules with identical molecular path codes, and it grows for less similar structures.

Figure 11.7 shows the computation of the dissimilarities among three monocyclic monoterpenes. Compounds A and Y are the least similar of the three pairs with a *D* value of 193. Compounds G and Y are the most similar with a *D* value of 26. In the original paper, 29 compounds were compared for degree of similarity. For all $\frac{1}{2}(29 \times 28) = 406$ comparisons, the smallest value of *D* reported was 3 and the largest was 306. Randić and Wilkins discussed possible uses of this dissimilarity measure based on molecular path counts in the areas of structure elucidation and structural chemistry.

Program PATH

Program PATH is a FORTRAN language routine that enumerates the atomic and molecular paths of a molecule represented as a connection table. The computation is done by subroutine PATHCT, which is a somewhat revised version of the routine published by Randić et al. (1979). The structure being considered is input through subroutine CTIS and is passed to PATHCT through the common block named CTI. The quantity of output generated is determined by the user's answer to a query. If less output is desired, the numbers of paths of each length for each atom are output. If more output is specified, the detailed enumeration of each and every path in the structure is also output.

The routine has been executed three times for example molecules of extremely different skeletons. The first compound is a six non-H atom unsaturated ether, which has 21 total paths, with the longest being two paths of length 4. The second compound is cubane with eight carbon atoms and 12 single bonds in a cubic arrangement. Because of the symmetry and highly connected nature of the structure, cubane has 452 total paths, with the longest being 72 paths of length 7. It also has 120 paths of length 5. The third compound is the simple alkane octane with eight carbons and just seven bonds. It contains a total of 36 paths, with the longest being the single path of length 7. The comparison of cubane and *n*-octane, each with eight carbons, but having vastly different numbers of paths, shows the potential power of path counts as means to characterize the skeletons of organic compounds.

11.5 TOPOLOGICAL INDICES

One long-standing goal of chemistry is to represent chemical structures in numerical form as succinctly but as completely as possible. When molecular

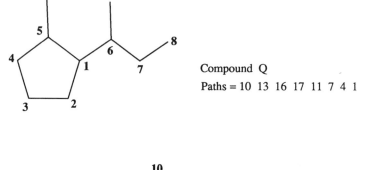

Compound Q
Paths = 10 13 16 17 11 7 4 1

Compound A
Paths = 10 12 10 8 7 7 6 2

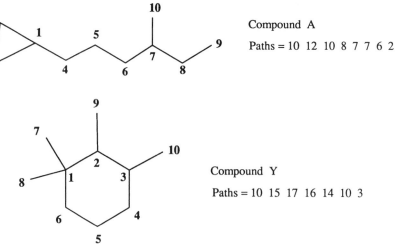

Compound Y
Paths = 10 15 17 16 14 10 3

Similarities: (QA) = 139

(QY) = 26

(AY) = 193

Figure 11.7 Dissimilarities among three monocyclic monoterprenes.

structures are represented as graphs, then this quest is equivalent to seeking ways to represent graphs as numbers. Topological indices have been developed by chemists in pursuit of this goal. A *topological index* is a numerical quantity that is mathematically derived in a direct and unambiguous manner from the structural graph of a molecule. Since isomorphic graphs possess identical values for any given topological index, these indices are referred to as *graph invariants*. Topological indices ordinarily encode both molecular size and shape at the same time. There have been more than 50 topological indices presented in the literature since their first development. In this section we will discuss several widely used topological indices and show how they have been applied to chemical problems.

In 1975, Randić proposed a topological index that has evolved into the most widely used in chemical studies (Randić 1975). This branching index was originally defined as

$$R = \sum_{\substack{\text{all} \\ \text{bonds}}} \frac{1}{(mn)^{1/2}}$$

where the summation includes one term for each edge in the hydrogen-suppressed structural graph. Thus, when the graph is representing an organic molecule, there is one term in the summation for each bond in the structure. The variables m and n are the degrees of the adjacent nodes joined by each edge. In chemical structure terms, this is the number of bonds attached to each atom participating in the bond.

Figure 11.8 shows the sequence of steps for the calculation of the value for

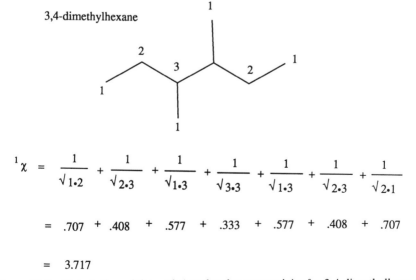

$$^1\chi = \frac{1}{\sqrt{1 \cdot 2}} + \frac{1}{\sqrt{2 \cdot 3}} + \frac{1}{\sqrt{1 \cdot 3}} + \frac{1}{\sqrt{3 \cdot 3}} + \frac{1}{\sqrt{1 \cdot 3}} + \frac{1}{\sqrt{2 \cdot 3}} + \frac{1}{\sqrt{2 \cdot 1}}$$

$$= .707 + .408 + .577 + .333 + .577 + .408 + .707$$

$$= 3.717$$

Figure 11.8 Calculation of the path 1 molecular connectivity for 3,4-dimethylhexane.

this simple topological index for the example molecule 3,4-dimethylhexane. At the top of the figure, the molecular structure is shown as a graph with the degree of each node labeled. The value $^1\chi = 3.717$ reflects both the size and branching of the structure. It is related to the size of the molecule because when extra atoms and bonds are added, more terms are added to the summation, and the value grows. $^1\chi$ is also related to the degree of branching of the molecule because when more branching occurs, then the denominators for those terms get larger and the terms themselves get smaller, thus decreasing the overall value for the index.

The normal valence of carbon atoms in organic compounds is 4, so the valences of the nodes in the hydrogen-suppressed structural graphs of simple alkanes are limited to the values of 1, 2, 3, and 4. Therefore, there are 10 possible sets of edges: 1–1, 1–2, 1–3, 1–4, 2–2, 2–3, 2–4, 3–3, 3–4, 4–4. The 1–1 edge type can occur only in ethane, and the edges of type 1–4 and 2–2 each yield the same product. Thus, this branching index is based on the decomposition of a compound into eight different carbon–carbon bond types. Since there are not very many different bond types, it follows that the branching index value can be the same for different molecules. For example, 3-methylheptane and 4-methylheptane have identical values for this branching index.

The simple branching index discussed above involves a summation over all paths of length one in the graph being treated. This viewpoint has been extended to include the definition of additional indices corresponding to paths of lengths 2, 3, or greater and to other subgraphs such as clusters and path clusters. This entire class of topological indices are commonly called *molecular connectivity indices*. The original Randić branching index is referred to as the *path 1 molecular connectivity* $^1\chi$. The higher-order indices are calculated by equations analogous to the simple equation for path 1 molecular connectivity.

The following equation generates the path 2 molecular connectivity from the degrees of the three edges involved in the definitions of paths of length 2:

$$^2\chi = \sum_{\substack{\text{length 2} \\ \text{paths}}} \frac{1}{(mnp)^{1/2}}$$

where m, n, and p are the degrees of the atoms of each path of length 2. For 3,4-dimethylhexane there are six terms in the summation for the six paths of length 2 in the molecule. The denominator contains the following terms: $(1 \cdot 2 \cdot 3)^{1/2}$, $(2 \cdot 3 \cdot 3)^{1/2}$, $(2 \cdot 3 \cdot 1)^{1/2}$, $(1 \cdot 3 \cdot 2)^{1/2}$, and $(3 \cdot 2 \cdot 1)^{1/2}$ and the overall value for $^2\chi$ for 3,4-dimethylhexane is 2.201.

The simplest molecular connectivity indices described to this point do not allow for the differentiation of atom types. To generalize the molecular connectivity indices and make them more useful for the characterization of organic molecules containing heteroatoms, the following enhancement has been developed. In the denominator of the defining equation, delta values were used in place of the degree of the node. The delta values are defined as

$$\delta^v = Z^v - h$$

TABLE 11.1 **Valence Delta Values for Carbon, Nitrogen, and Oxygen in Different Bonding Environments**

$-CH_3$	1	$-NH_2$	3	$-OH$	5
$-CH_2-$	2	$>NH$	4	$-O-$	6
$>CH-$	3	$=NH$	4	$=O$	6
$=CH-$	3	$>N-$	5		
$=C<$	4	$=N-$	5		
$>C<$	4	$\equiv N$	5		

where Z^v is the number of valence electrons for the atom, and h is the number of attached hydrogens. Thus, a carbonyl oxygen has a $\delta^v = 6$ and a nitrogen atom as a secondary amine has a value of $\delta^v = 4$. Table 11-1 provides a complete list of the valence delta values for carbon, nitrogen, and oxygen atoms in various bonding environments.

Molecular connectivity indices calculated with these delta values are called *valence molecular connectivity* and are given a superscript "v." Figure 11.9 shows the calculation of the valence path 1 molecular connectivity index $^1\chi^v$ for 2-(methylamino)-propionic acid methyl ester (or *N*-methylalanine methyl ester).

The valence molecular connectivity index has been correlated with many

2-(methylamino)-propionic acid methyl ester

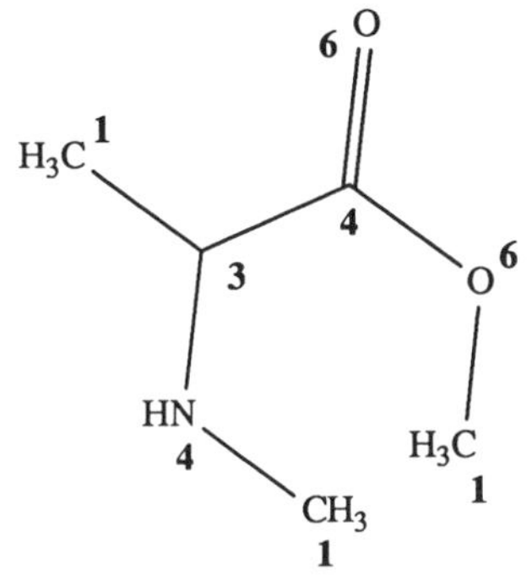

$$^1\chi^{\,v} = \frac{1}{\sqrt{1\cdot3}} + \frac{1}{\sqrt{3\cdot4}} + \frac{1}{\sqrt{4\cdot1}} + \frac{1}{\sqrt{3\cdot4}} + \frac{1}{\sqrt{4\cdot6}} + \frac{1}{\sqrt{4\cdot6}} + \frac{1}{\sqrt{6\cdot1}}$$

$$= 2.47$$

Figure 11.9 Calculation of the valence path 1 molecular connectivity for 2-(methyl-amino)propionic acid methyl ester.

physicochemical properties of organic compounds. The index is easy to compute and thus is more accessible than values derived from complicated experimental measurements. An example is the demonstrated correlation between the log P and $^1\chi^v$ for 138 simple organic compounds, including 24 esters, 9 carboxylic acids, 49 alcohols, 28 amines, 16 ketones, and 12 ethers (Murray et al. 1975). The log P for a compound is the logarithm of the partition coefficient of the compound between water and 1-octanol. Log P of organic compounds has been shown in hundreds of studies to be related to biological activity and environmental transport rates, and it is thus of very great interest.

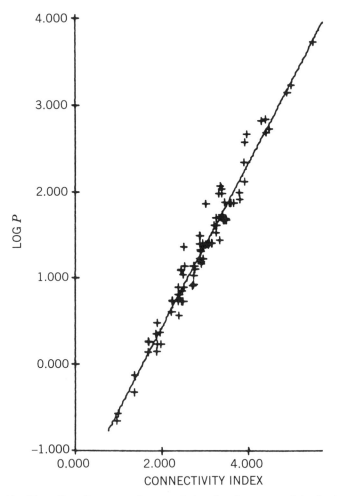

Figure 11.10 Plot of log P versus valence path 1 molecular connectivity for 138 simple organic compounds. [Reprinted with permission from *Journal of Pharmaceutical Sciences,* **64**, 1978 (1975).]

Log P is ordinarily measured by the shake-flask method or by chromatographic methods, either of which is time-consuming and expensive. The correlation between measured log P values and calculated $^1\chi^v$ values is shown visually in Figure 11.10. The equation of the fit line is

$$\log P = 0.95\,^1\chi^v - 1.48$$

$$n = 138 \qquad r = 0.986 \qquad s = 0.152$$

Thus, molecular connectivity indices can be used to encode information about molecular structures that are also represented by experimentally measured quantities. The molecular connectivity indices are the most widely used topological indices for quantitative structure–activity relationship and quantitative structure-property relationship studies. Kier and Hall (1986) include a list of 158 journal references in which molecular connectivity played a prominent role. Papers continue to appear in which molecular connectivity indices play a prominent role.

```
      PROGRAM PATH
C....
C.... PROGRAM TO ENUMERATE ATOMIC AND MOLECULAR PATHS
C....
      DIMENSION LB(32),LC(32,32)
      COMMON /CTI/ NATOMS,NBONDS,IFA(32),ITA(32),NBTYPE(32),NATYPE(32)
      COMMON /IOUNIT/ NINP,NOUT
      NINP=5
      NOUT=6
      WRITE (NOUT,1)
    1 FORMAT (' PATH PROGRAM',/)
C....
      WRITE (NOUT,3)
    3 FORMAT (' ENTER OUTPUT LEVEL:',/,5X,'1=LEAST, 2=MOST')
      READ (NINP,*) IPR
C....
   12 CALL CTIS
C....
      DO 25 I=1,NATOMS
      DO 25 J=1,NATOMS
   25 LC(I,J)=0
      DO 27 J=1,NBONDS
      LC(IFA(J),ITA(J)) = 1
   27 LC(ITA(J),IFA(J)) = 1
      NSIZE=NATOMS
      CALL PATHCT (LB,LC,NSIZE,IPR)
      STOP
      END
C-----------------------------------------------------
      SUBROUTINE PATHCT (LB,LC,NATM,IPR)
C....
      IMPLICIT INTEGER (A-Z)
      DIMENSION A(32),LB(32),LC(32,32),R(32),S(32),P(32),X(32)
      COMMON /IOUNIT/ NINP,NOUT
C....
C.... A array contains the atom-based path count vector
C.... LB array contains the molecule-based path count vector
```

```
C.... LC array conatins the connection table
C.... R(I) contains the previous atom in current path
C.... S(I) contains the current column in the
C....       condensed C matrix of atom i
C.... IPR controls output
C....       2 for output of all paths
C....       1 for output of atomic path counts
C....       0 for none
C....
      DATA P/' 1',' 2',' 3',' 4',' 5',' 6',' 7',' 8',' 9','10','11',
     X '12','13','14','15','16','17','18','19','20','21','22','23',
     X '24','25','26','27','28','29','30','31','32'/
      DATA BLANK,X/33*' '/,NUMMAX/32/
      IF (IPR.GT.0) WRITE (NOUT,1000)
      DO 30 I=1,NATM
      K=1
      DO 20 J=1,NATM
      IF (LC(I,J).EQ.0) GO TO 20
      LC(I,K)=J
      IF (J.EQ.K) GO TO 10
      LC(I,J)=0
   10 K=K+1
   20 CONTINUE
   30 CONTINUE
      DO 34 J=1,NUMMAX
   34 LB(J)=0
      DO 100 KK=1,NATM
      IF (IPR.NE.0) WRITE (NOUT,1070) KK
      DO 35 J=1,NUMMAX
      A(J)=0
      S(J)=0
   35 R(J)=0
      N=1
      X(N)=P(KK)
      R(KK)=-1
      A(1)=1
      K=KK
   38 S(K)=S(K)+1
      T=S(K)
      T=LC(K,T)
      IF (T.EQ.0) GO TO 50
      IF (R(T).NE.0) GO TO 38
      R(T)=K
      K=T
      N=N+1
      IF (IPR.EQ.0) GO TO 42
      X(N)=P(K)
   42 FLG=1
      A(N)=A(N)+1
      GO TO 38
   50 IF (FLG.EQ.0) GO TO 54
      IF (IPR.LT.2) GO TO 53
      WRITE (NOUT,1030) (X(LL),LL=1,N)
   53 FLG=0
   54 DO 60 LL=1,N
   60 X(LL)=BLANK
      N=N-1
      IF (N.LT.0) GO TO 68
      S(K)=0
      T=R(K)
      R(K)=0·
      K=T
      IF (K.GT.0) GO TO 38
   68 DO 70 LL=1,NATM
```

```
   70 LB(LL)=LB(LL)+A(LL)
      IF (IPR.NE.0) WRITE (NOUT,1040) (A(LL),LL=1,NATM)
  100 CONTINUE
      TOTAL=LB(1)
      DO 110 LL=2,NATM
      LB(LL)=LB(LL)/2
  110 TOTAL=TOTAL+LB(LL)
      IF (IPR.EQ.0) GO TO 111
      WRITE (NOUT,1050) (LB(LL),LL=1,NATM)
      WRITE (NOUT,1090) TOTAL
  111 RETURN
 1000 FORMAT (/,' PATH ENUMERATION ROUTINE',/)
 1030 FORMAT (' ',25('    ',A2))
 1040 FORMAT (' ',20I3)
 1050 FORMAT (/,' MOLECULAR PATH COUNTS',/,' ',25I3)
 1070 FORMAT (' PATHS FOR ATOM',I3)
 1090 FORMAT (/,' NUMBER OF PATHS =',I7)
      END
C-------------------------------------------------------------
      SUBROUTINE CTIS
C....
C.... CONNECTION TABLE INPUT SUBROUTINE
C....
      COMMON /CTI/ NATOMS,NBONDS,IFA(32),ITA(32),NBTYPE(32),NATYPE(32)
      COMMON /IOUNIT/ NINP,NOUT
C....
      WRITE (NOUT,1)
    1 FORMAT (' CONNECTION TABLE INPUT',/)
      WRITE (NOUT,2)
    2 FORMAT (' INPUT NATOMS, NBONDS')
C.... INPUT NUMBER OF ATOMS AND NUMBER OF BONDS
      READ (NINP,*) NATOMS,NBONDS
      WRITE (NOUT,3)
    3 FORMAT (' INPUT ATOMS TYPES: 1=C, 2=O, 3=N, 4=S, 5=CL')
C.... INPUT ATOM TYPES
      READ (NINP,*) (NATYPE(K),K=1,NATOMS)
      WRITE (NOUT,4)
    4 FORMAT (' INPUT TRIPLES: (FROM ATOM, TO ATOM, BOND TYPE)')
C.... INPUT BONDING INFORMATION
      READ (NINP,*) (IFA(K),ITA(K),NBTYPE(K),K=1,NBONDS)
      RETURN
      END
```

```
PATH PROGRAM

ENTER OUTPUT LEVEL:
    1=LEAST, 2=MOST
1
 CONNECTION TABLE INPUT

 INPUT NATOMS, NBONDS
6 5
 INPUT ATOMS TYPES: 1=C, 2=O, 3=N, 4=S, 5=CL
1 1 1 2 1 1
 INPUT TRIPLES: (FROM ATOM, TO ATOM, BOND TYPE)
1 2 1   2 3 2   2 6 1   3 4 1   4 5 1

 PATH ENUMERATION ROUTINE

 PATHS FOR ATOM  1
   1  1  2  1  1  0
 PATHS FOR ATOM  2
   1  3  1  1  0  0
```

```
PATHS FOR ATOM  3
  1  2  3  0  0  0
PATHS FOR ATOM  4
  1  2  1  2  0  0
PATHS FOR ATOM  5
  1  1  1  1  2  0
PATHS FOR ATOM  6
  1  1  2  1  1  0

MOLECULAR PATH COUNTS
  6  5  5  3  2  0

NUMBER OF PATHS =     21


 PATH PROGRAM

 ENTER OUTPUT LEVEL:
    1=LEAST, 2=MOST
1
 CONNECTION TABLE INPUT

 INPUT NATOMS, NBONDS
8 12
 INPUT ATOMS TYPES: 1=C, 2=O, 3=N, 4=S, 5=CL
1 1 1 1 1 1 1 1
 INPUT TRIPLES: (FROM ATOM, TO ATOM, BOND TYPE)
1 2 1   1 4 1   1 5 1
2 3 1   2 6 1
3 4 1   3 7 1
4 8 1
5 6 1   5 8 1
6 7 1
7 8 1


 PATH ENUMERATION ROUTINE

PATHS FOR ATOM  1
  1   3   6 12 18 30 24 18
PATHS FOR ATOM  2
  1   3   6 12 18 30 24 18
PATHS FOR ATOM  3
  1   3   6 12 18 30 24 18
PATHS FOR ATOM  4
  1   3   6 12 18 30 24 18
PATHS FOR ATOM  5
  1   3   6 12 18 30 24 18
PATHS FOR ATOM  6
  1   3   6 12 18 30 24 18
PATHS FOR ATOM  7
  1   3   6 12 18 30 24 18
PATHS FOR ATOM  8
  1   3   6 12 18 30 24 18

MOLECULAR PATH COUNTS
  8 12 24 48 72120 96 72

NUMBER OF PATHS =    452
```

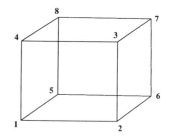

```
PATH PROGRAM

ENTER OUTPUT LEVEL:
   1=LEAST, 2=MOST
1
CONNECTION TABLE INPUT

INPUT NATOMS, NBONDS
8 7
INPUT ATOMS TYPES: 1=C, 2=O, 3=N, 4=S, 5=CL
1 1 1 1 1 1 1 1
INPUT TRIPLES: (FROM ATOM, TO ATOM, BOND TYPE)
1 2 1   2 3 1   3 4 1   4 5 1   5 6 1   6 7 1   7 8 1
PATH ENUMERATION ROUTINE

PATHS FOR ATOM  1
   1  1  1  1  1  1  1  1
PATHS FOR ATOM  2
   1  2  1  1  1  1  1  0
PATHS FOR ATOM  3
   1  2  2  1  1  1  0  0
PATHS FOR ATOM  4
   1  2  2  2  1  0  0  0
PATHS FOR ATOM  5
   1  2  2  2  1  0  0  0
PATHS FOR ATOM  6
   1  2  2  1  1  1  0  0
PATHS FOR ATOM  7
   1  2  1  1  1  1  1  0
PATHS FOR ATOM  8
   1  1  1  1  1  1  1  1

MOLECULAR PATH COUNTS
   8  7  6  5  4  3  2  1

NUMBER OF PATHS =      36
```

REFERENCES

Balaban, A. T., "Applications of Graph Theory in Chemistry," *J. Chem. Inf. Comput. Sci.*, **25**, 334–343 (1985).

Balaban, A. T. (ed.), *Chemical Applications of Graph Theory,* Academic Press, New York, 1976.

Cayley, A., *Philos. Mag.*, **67**, 444 (1874).

Essam, J. W., and M. E. Fisher, "Some Basic Definitions in Graph Theory," *Rev. Mod. Phys.*, **42**, 272–288 (1970).

Grant, D. M., and E. G. Paul, "Carbon-13 Magnetic Resonance. 11. Chemical Shift Data for the Alkanes," *J. Am. Chem. Soc.*, **86**, 2984–2990 (1964).

Hansen, P. J., and P. C. Jurs, "Chemical Applications of Graph Theory. Part I. Fundamentals and Topological Indices," *J. Chem. Ed.*, **65**, 574–580 (1988).

Hansen, P. J., and P. C. Jurs, "Chemical Applications of Graph Theory. Part II. Isomer Enumeration," *J. Chem. Ed.*, **65**, 661–664 (1988).

Harary, F., *Graph Theory,* Addison-Wesley, Reading, MA, 1969.

Harary, F., E. M. Palmer, R. W. Robinson, and R. C. Read, "Polya's Contributions to Chemical Enumeration," in *Chemical Applications of Graph Theory,* A. T. Balaban (ed.), Academic Press, New York, 1976.

Henze, H. R., and C. M. Blair, *J. Am. Chem. Soc.,* **53**, 3042 (1931); **53**, 3077 (1931); **54**, 1098 (1932); **54**, 1538 (1932).

Kennedy, J. W., "Small Graphs, Graph Theory and Chemistry," in *Data Processing in Chemistry,* Z. Hippe (ed.), Elsevier Scientific, Amsterdam, 1981.

Kier, L. B., and L. H. Hall, *Molecular Connectivity in Chemistry and Drug Research,* Academic Press, New York, 1986.

Lederberg, J., G. L. Sutherland, B. G. Buchanan, E. A. Feigenbaum, A. V. Robertson, A. M. Duffield, and C. Djerassi, "Applications of Artificial Intelligence for Chemical Inference. 1. The Numbers of Possible Organic Compounds, Acyclic Structures Containing C, H, O, and N," *J. Am. Chem. Soc.,* **91**, 2973–2976 (1969).

Lindeman, L. P., and J. Q. Adams, "Carbon-13 Nuclear Magnetic Resonance Spectroscopy. Chemical Shifts for the Paraffins through C9," *Anal. Chem.* **43**, 1245–1252 (1971).

Lynch, M. F., J. M. Harrison, and W. G. Town, *Computer Handling of Chemical Structure Information,* Macdonald, London, 1971.

Morgan, H. L., "Generation of a Unique Machine Description for Chemical Structures: A Technique Developed at Chemical Abstracts Service," *J. Chem. Soc.,* **5**, 107–113 (1965).

Murray, W. J., L. H. Hall, and L. B. Kier, "Molecular Connectivity III: Relationship to Partition Coefficients," *J. Pharm. Sci.,* **64**, 1978 (1975).

Nourse, J. G., D. H. Smith, R. E. Carhart, and C. Djerassi, "Applications of Artificial Intelligence for Chemical Inference. 33. Computer-Assisted Elucidation of Molecular Structure with Stereochemistry," *J. Am. Chem. Soc.,* **102**, 6289–6295 (1980).

O'Korn, L. J., "Algorithms in the Computer Handling of Chemical Information," in *Algorithms for Chemical Computations,* R. E. Christoffersen (ed.), American Chemical Society, Washington, DC, 1977.

Polya, G., "Kombinatorische Anzahlbestimmungen fur Gruppen, Graphen und Chemische Verbindungen," *Acta Math.,* **68**, 145–254 (1937).

Randić, M., "On Characterization of Molecular Branching," *J. Amer. Chem. Soc.,* **97**, 6609–6615 (1975).

Randić, M., "Characterization of Atoms, Molecules, and Classes of Molecules Based on Path Enumerations," *MATCH,* **7**, 5–64 (1979).

Randić, M., and C. L. Wilkins, "Graph Theoretical Approach to Recognition of Structural Similarity in Molecules," *J. Chem. Inf. Comput. Sci.,* **19**, 31–37 (1979).

Randić, M., G. M. Brissey, R. B. Spencer, and C. L. Wilkins, "Search for All Self-Avoiding Paths for Molecular Graphs," *Comput. Chem.,* **3**, 5–13 (1979).

Randić, M., G. M. Brissey, R. B. Spencer, and C. L. Wilkins, "Use of Self-Avoiding Paths for Characterization of Molecular Graphs with Multiple Bonds, *Comput. Chem.* **4**, 27–43 (1980).

Rouvray, D. H., "Isomer Enumeration Methods," *Chem. Soc. Rev.,* **3**, 355–372 (1974).

Rouvray, D. H., "Predicting Chemistry from Topology," *Sci. Am.,* **255**, 40–47 (1986).

Trinajstić, N., *Chemical Graph Theory,* 2nd ed., CRC Press, Boca Raton, FL, 1992.

Wipke, W. T., S. R. Heller, R. J. Feldmann, and E. Hyde (eds.), *Computer Representation and Manipulation of Chemical Information,* Wiley-Interscience, New York, 1974.

Wipke, W. T., and T. M. Dyott, "Stereochemically Unique Naming Algorithm," *J. Am. Chem. Soc.,* **96**, 4834–4842 (1974).

12

SUBSTRUCTURE SEARCHING

12.1 PRINCIPLES OF SUBSTRUCTURE SEARCHING

Chemical information systems are the repositories for chemical structures and associated information that has been collected and archived. Once an effective method has been developed for storing chemical structures in a computer system and has been used to generate a file of chemical structures, the next step is to consider the uses for such a file.

One reason for the existence of a chemical information system is the ability to retrieve useful information from it. Questions such as the following might prompt a chemist to utilize a chemical information system:

Has my query compound ever been reported?
Show me the compounds in the file that are most structurally similar to my query compound.
What properties are known for my query compound?
How can my query compound be synthesized?
What compounds have substructural features similar to those of my query compound?

Most of these questions are essentially structural, that is, the heart of the question is a structure or a partial structure. The ability to carry out chemical substructure searching is an important attribute of chemical information systems.

The essence of substructure searching is simple: identifying all the compounds in a given file that contain the specified substructural feature. Substructure searching is a special case of the more general problem of subgraph isomorphism, which involves the determination of whether one graph is a subgraph of another graph. One is determining whether one graph is embedded within another graph. For one graph (substructure) to be a subgraph of another graph (structure), there must be a mapping of all the nodes (atoms) and edges (bonds) from the larger structure to the smaller. This matching problem is one member of the class of problems known as *NP-complete*. This means that in the worst case, the time required to perform such a match will be exponentially related to the number of atoms in the graph being processed. This is because a brute-force matching of a substructure with n_{ss} atoms with a structure of n_s atoms could require $n_s!/(n_s + n_{ss})!$ mappings to be compared, a number that becomes quite impractical for even moderate numbers of atoms in the structures being compared. Although the average time required to answer the query is of more interest than the worst-case possibility, brute-force approaches are not effective for substructure searching, and other methods have been developed.

The effectiveness of procedures to accomplish substructure searching depends on many factors, such as the following:

1. The method used for representation of the structures in the file being searched
2. The method used for representation of the query substructural features
3. The sophistication of the algorithm employed
4. The size of the structure file being searched

We will investigate the effect of each of these factors in the following pages.

Two measures of value for substructure search systems are precision and recall. *Recall* is the fraction of the matches that should be found that were found. If a substructure search yields all the structures that it should have found, recall is 100%. If a substructure search yields a hit list that contains half of the structures that should have been found, recall is 50%. *Precision* is the fraction of the retrieved structures that are relevant. If a substructure search yields 50 structures but only 25 of them are relevant, precision is 50%. There is usually an inverse relationship between recall and precision. In other words, when working with algorithms for substructure search—which ordinarily have adjustable parameters and other possible variations—one observes that as one of these measures goes up, the other goes down. A substructure search can be adjusted to capture all of relevant structures (i.e., 100% recall), but at the cost of retrieving some irrelevant structures. A substructure search can be adjusted to retrieve almost no irrelevant structures, but at the cost of failing to retrieve all relevant structures.

Substructure searching of files of structures is a computationally complex and expensive task. As mentioned above, it is an NP-complete problem, so the time expended to perform a substructure search can grow exponentially in worst-case scenarios. However, the time required for ordinary searches increases rapidly with the size of the problem. This leads to the necessity of breaking down substructural searching into component tasks.

Substructure searching systems fall into two classes of algorithms. The first type of system to be developed uses a two-step procedure that involves a first step of fragment screening followed by atom-by-atom search. More recently, with the advent of cheap, large computer memory, an alternative type of approach has been developed that is more efficient during search.

Atom-by-Atom Search Algorithms with Fragment Screening

The individual components of the overall substructure searching task using a fragment screening and atom-by-atom search are shown in Figure 12.1. The structural entries that are potential matches for the query substructure are sifted out of the main structure file, and only then are they searched in complete detail. The initial sifting is done with fragment screens. This simple

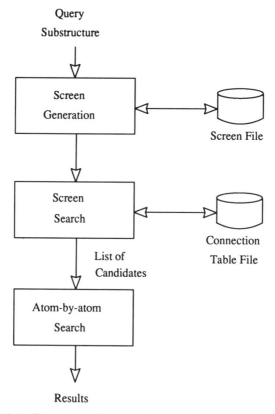

Figure 12.1   Flowchart of the steps in a typical substructure search.

and rapid sifting method eliminates the vast majority of the structures from further consideration. The screens represent a set of structural characteristics that can be identified in both the substructures and the structures being searched. The screen search of the structure file picks out all those structures that match the screens of the query substructure. The cost and response time of an on-line substructure searching system are determined to a large extent by the following two factors.

1. The number of structures from the file being searched that are passed by the screen search
2. The number of actual valid answers to the substructural query posed

Thus, the design and use of effective screens is extremely important to the overall effectiveness of a substructure search system.

Screen Generation   Screen searching is implemented by defining a set of structural features that characterize the structures in the file and the substructural query. The screens are derived from the connection-table representations of the structures in the file just once, and they are stored in a screen file. The screens for the query substructure are derived as the first step in a substructure search. The objective in defining a set of screens is to encode as much structural information about the relevant structure or substructure as possible. However, the screen set must not be unduly long or computationally expensive to determine.

Simple screens can be derived directly from the connection table of a structure because the connection table contains information about the atoms, bonds, and connections present in the compound. However, simple screens are not sufficient to screen out a sufficiently large fraction of the compounds from the structure file during searching.

A much more effective set of screens can be derived by the use of chemically significant fragments as screens. Such fragments can also be generated from the connection tables by software. The set of chemically significant screens to be used in a particular system can be determined by algorithm. The molecules of the structure file are fragmented by a given set of rules. Thus, the fragments generated reflect the characteristics of the structures in the file, and both the number and type of fragments generated are characteristic of the file.

An example of a set of easily determined information that might constitute a screen set is shown in Figure 12.2. Atom type, bond type, and ring counts are simple screens that characterize a molecular structure to some degree. Thus, if a substructure containing two rings, or bromine, or nitrogen were being matched against a hydrocarbon containing only carbon and hydrogen, a check of such simple screens would eliminate the hydrocarbon from further consideration. Screens can also note the presence or absence of small simple substructures such as those shown in Figure 12.2. If a potential candidate

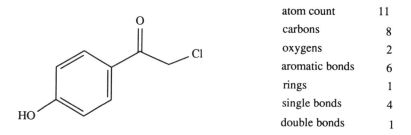

atom count	11
carbons	8
oxygens	2
aromatic bonds	6
rings	1
single bonds	4
double bonds	1

Possible substructure screens:

Figure 12.2 Example molecular structure and examples of relevant screens.

structure lacks one of the necessary screens of the query substructure, it can be eliminated from further consideration. Evidently, the intent in screening is to achieve 100% recall with 100% precision. If this were feasible, there would be no need for the further refinement of the searching procedure, the atom-by-atom iterative search. Since screening does not produce the final answer to a substructural query, but rather produces a hit list of molecular structures that pass screening, the atom-by-atom search is necessary for final confirmation of a hit.

If the individual pieces of structural information can be reduced to bits, an entire screen set can be stored in a binary representation. This leads to very great efficiency since the screen set for a candidate structure and the screen set for the query substructure can then be compared using logical functions. This is, in fact, the common practice in actual searching systems in order to make screening as efficient as possible.

For files of compounds of even moderate size, the total number of possible screens is very large. Accordingly, some criteria must be adopted for deciding which screens to use. One important criterion is that, if possible, each screen should split the file of compounds in half. That is, half of the compounds in the file should have a value for the screen of 0, and half should have the value of 1. Information theory tells us that this would allow the maximum amount of information to be stored in the set of screens. Of course, in practice the screens generated from a file of structures will not have this ideal characteristic, but the goal remains valid.

A great deal of information can be coded in binary form. Screens can be designed to code nearly any structural information in this form. As examples, consider the following potential screens:

Screen for carbon number:

    0 if carbon number is less than the mean

    1 of carbon number is greater than the mean

Screen for nitrogen presence:

    0 if no N present

    1 if at least one N present

Screen for oxygen presence:

    0 if no O present

    1 if at least one O present

Screen for molecular size:

    0 if the molecular weight is less than the mean

    1 if the molecular weight is greater than the mean

Screen for the carbonyl substructure:

    0 if carbonyl moiety absent

    1 if present

Screen for aromaticity:

    0 if no aromatic bonds present

    1 if aromatic bonds are present

Screen for five-member ring presence:

    0 if no five-member rings are present

    1 if at least one five-membered ring is present

Screen for benzene ring presence:

    0 if no six-member aromatic rings present

    1 if at least one six-member aromatic ring present

Screen for presence of multiple rings:

    0 if one or fewer rings present

    1 if two or more rings present

Note that the definitions of some of these screens are derived from the properties of the entire structural file that is to be searched. Thus, the screens are custom-designed to sort as efficiently as possible through the file. In the screen examples listed above, the carbon-number example and the molecular-weight example show this property of tailoring to the set of structures.

Screen generation algorithms much more elaborate and efficient to use have been devised. For example, Willett (1979) has described a screen set generation method that develops screens of equal frequency that are structural and are very efficient to employ. The Chemical Abstracts Service, which must deal with a file of structures in the millions, has devised an extremely elaborate screen set with more than 6000 different screen definitions (Dittmar et al. 1983).

The CAS ONLINE search system has a dictionary of screens that are used by the system. The screens have good selectivity as well chemical significance. There are 12 types of screens employed in three classes: (1) augmented atom screens that describe atoms and their nearest neighbors and include augmented atom (AA), hydrogen augmented atom (HA), and twin augmented atom (TW) screens; (2) linear sequence screens that describe linear strings of atoms (excluding hydrogens) and are atom sequence (AS), bond sequence (BS), and connectivity sequence (CS) screens; and (3) general structural feature screens that include ring count (RC), type of ring (TR), atom count (AC), degree of connectivity (DC), element composition (EC), and graph modifier (GM) screens. Augmented atom screens are descriptions of atoms and their nearest-neighbor attachments (excluding hydrogens), for example. Atom sequence screens are descriptions of linear sequences of four, five, or six atoms (excluding hydrogen). A screen set of this complexity and sophistication is necessary for CAS ONLINE because of the size of the file being searched.

Atom-by-Atom Search   After the screening operation is finished, structures that are potential matches to the substructural query are listed. These potential hits have all passed the screening step. The field of structures can be further narrowed by detailed atom-by-atom comparison of the substructural query with each potential hit.

The matching of a substructure against a structure by an atom-by-atom search is, as mentioned above, a subgraph isomorphism problem. It is simple in concept but complicated and time-consuming in practice. The method begins with the least commonly occurring atom in the substructure and looks through the structure for an equivalent atom. When an equivalent atom is found, the next nearest-neighbor atoms of the starting atom are checked for equivalence. When equivalences are found, the next nearest atoms are checked. Whenever an equivalence is not found, the algorithm backs up to the last successful checking point. The algorithm exhaustively tests all such correspondences between the atoms of the substructure query and the structure until one of two conditions is met, either (1) a match is found or (2) all possible correspondences have been checked.

Consider a specific example of an atom-by-atom search as shown in Figure 12.3. The structure shown is to be examined to determine whether the substructure is present. Say that the search starts with the carbonyl carbon of the substructure, atom 1 in the substructure. The atoms of the structure would be searched, starting with atom 1, until a carbon atom is found, in this case atom 1 in the structure. Then the next nearest neighbors of substructure atom 1 would be compared to the next nearest-neighbor atoms of atom 1 in the structure. Atom 2 in the substructure is equivalent to atom 4 in the structure, and the bond between atoms 1 and 2 in the substructure is double, as is the bond between atoms 1 and 4 in the structure. Atom 3 of the substructure is then compared to atom 2 in the structure, and they are found to be different. The algorithm backtracks to the last correct equivalence, and it next checks atoms of the substructure to atom 3 of the structure. Equivalence is found. Finally, atom 4 of the substructure is compared to atom 6 of the structure, and they are found to be equivalent. All the atoms of the substructure have been matched against atoms in the structure, and in the proper bonding configuration, the substructure has thus been found in its entirety. The correspondances of atoms in the structure and the substructure are shown in the lower part of Figure 12.3. The search of this structure is complete, and the algorithm can now proceed to the next structure in the file.

The amount of backtracking that is necessary in an atom-by-atom search depends on the amount of symmetry in the structure, the nature of the substructure, the numbering of the structure and substructure, and many other factors. The overall speed of the search depends on the amount of backtracking as well as the overall size of the substructure. The potential complexity of the atom-by-atom search makes it evident why efficient screening is necessary to ensure the smallest possible number of structures to be handled by atom-by-atom search.

Direct Algorithms without Fragment Screening

The availability of large computer memories and even larger disk and CD-ROM (compact-disk–read-only memory) storage has led to the development of completely new approaches to substructure searching. In these approaches, a great deal of complex preprocessing is done to set up files that can then be used quickly and efficiently during the actual substructure searching operation.

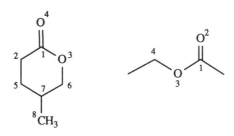

Atom Numbers Structure	Atom Numbers Substructure
1	1
4	2
3	3
6	4

Figure 12.3 Example structure and substructure being sought by atom-by-atom search.

Several systems using tree-structured files have been implemented. The molecular structures in the database to be stored are broken down into fragments and sorted into a hierarchy that takes into account the neighboring atom types, bonding patterns, and other factors. A complex hierarchical tree of search files is generated, and the use of these files leads to rapid searches. The need for the time-consuming atom-by-atom searches is eliminated. In the substructure search system (S4) (Hicks et al. 1990) the search tree is stored on CD-ROM, which permits the development and storage of an extremely large search tree to handle a very large database.

The development of tree-structured searches during the late 1980s led to the capability of searches on very large databases to be done quickly on desktop computers. Thus, the use of substructure searching has moved from large, central computer facilities to the immediate control of individual users, which makes access to this technology much more widely available.

Query Input and Output of Results   Chemical information systems that support substructure searching also support graphical input and output. Graphical description of molecular structures and substructures is the common language of chemistry, so it follows that users of chemical information systems would want to express their queries graphically and receive their answers graphically. Essentially all working chemical information systems that support substructure searching also support graphical input and display. These two tasks are completely dependent on the hardware available as well as on the software being used.

The purpose of graphical input of substructures is to let the chemist user express a query in its natural language. The software can translate the query into the connection-table storage required to proceed with the search. Graphical input systems have been built with light pens, cursor control, mouses, and graphics tables. In each case, the software is designed to mimic the motions of drawing a structure on a blackboard. There is, however, at least one substantial advantage in computer software drawing—previous structures can be recalled to the screen for modification. Each new structure or substructure need not be drawn from scratch.

The output of the results of searching can be as simple or sophisticated as the software designer wishes and as the hardware available allows. The usual output is the familiar sticklike structural drawing, once again, similar to blackboard drawings. The types of graphical displays available for presentation of chemical structures are discussed in Chapters 17 and 18.

12.2   A WORKING SYSTEM: CAS ONLINE

The Chemical Abstracts Service, an arm of the American Chemical Society, has been collecting abstracts of the chemical literature since the late nineteenth century. The process has been computerized since 1965, with all the stored information accessible to computer search through CAS ONLINE more recently. The Chemical Registry System files contain more than 10 million substances.

One popular type of search query involves the definition of the search query in terms of a structural diagram. The diagram can be defined by the user on a graphics terminal or an alphanumeric terminal. The exact way in which the query is input is dependent on the hardware available to the user. The CAS ONLINE system can interact with many of the popular graphical and alphanumeric terminals. A query to the system can consist of one structural diagram or several structural diagrams. The Boolean logical operators AND, OR, and NOT can be used in conjunction with the query parts. Screens can also be constructed by the user; these are features that must be present in the sought structures in addition to the structural diagrams.

The CAS ONLINE system sifts the file being searched with the screens supplied by the user in addition to its own screens. It then applies an iterative atom-by-atom search to those structures passing the screening process. The user can request the CAS registry numbers, full structural diagrams, and other information about any of the substances that are retrieved by the search.

REFERENCES

Ash, J. E., P. A. Chubb, S. E. Ward, S. M. Welford, and P. Willett, *Communication, Storage, and Retrieval of Chemical Information,* Wiley, New York, 1985.

Barnard, J. M., "Substructure Searching Methods: Old and New," *J. Chem. Inf. Comput. Sci.,* **33**, 532–538 (1993).

Dittmar, P. G., N. A. Farmer, W. Fisanick, R. C. Haines, and J. Mockus, "The CAS ONLINE Search System. 1. General System Design and Selection, Generation and Use of Search Screens," *J. Chem. Inf. Comput. Sci.,* **23**, 93–102 (1983).

Hicks, M. G., C. Jochum, and H. Maier, "Substructure Search Systems for Large Chemical Data Bases," *Anal. Chim. Acta,* **235**, 87–92 (1990).

Hicks, M. G., and C. Jochum, "Substructure Search Systems. 1. Performance Comparison of the MACSS, DARC, HTSS, CAS Registry MVSSS, and S4 Substructure Search Systems," *J. Chem. Inf. Comput. Sci.,* **30**, 191–199 (1990).

Lynch, M. E., J. M. Harrison, and W. G. Town, *Computer Handling of Chemical Structure Information,* MacDonald, London, 1971.

Lynch, M. F., J. M. Barnard, and S. W. Welford, "Generic Structure Storage and Retrieval," *J. Chem. Inf. Comput. Sci.,* **25**, 264–270 (1985).

Stobaugh, R. E., "Chemical Substructure Searching," *J. Chem. Inf. Comput. Sci.,* **25**, 271–275 (1985).

Tarjan, R. E., "Graph Algorithms in Chemical Computation," *A.C.S. Symp. Ser.,* **46**, 1–20 (1977).

Willett, P., "A Screen Set Generation Algorithm," *J. Chem. Inf. Comput. Sci.,* **19**, 159–162 (1979).

Willet, P., "A Review of Chemical Structure Retrieval Systems," *J. Chemometrics,* **1**, 139–155 (1987).

Wipke, W. T., S. R. Heller, R. J. Feldmann, and E. Hyde, *Computer Representation and Manipulation of Chemical Information,* Wiley, New York, 1974.

13

MOLECULAR MECHANICS AND MOLECULAR DYNAMICS

13.1 INTRODUCTION

Once the topology of a chemical structure is specified, the next level of information is the conformational or geometric properties of the structure. The three-dimensional structure of a molecule is a major determinant of its physical, chemical, and biological properties. To understand molecular properties and how molecules participate in biological reactions requires a knowledge of the three-dimensional structure of molecules. Chemists commonly use mechanical models of chemical structures as an aid to visualization. Looking at molecules as three-dimensional objects naturally leads to consideration of strain energies, that is, the relationship between conformation and the energetic requirements to adopt a particular geometry. In the past, chemists have relied largely on mechanical models to facilitate visualization, but these models have limitations that can have undesirable effects on one's reasoning. Now, methods for calculating molecular geometries and their associated energies are available, and they can be used as a routine research and teaching tool.

The method of molecular mechanics was introduced by Westheimer (1956) and refined by Wiberg (1965). It is known as the *strain energy minimization technique,* the *molecular mechanics method,* or the *force-field method.* This chapter describes this important area of chemical computation.

Basis of the Method

The fundamental assumption of molecular mechanics is that data determined experimentally for small molecules (bond lengths, bond angles, etc.) can be extrapolated to larger molecules. A molecule is considered to be a collection of atoms held together by simple elastic or harmonic forces derived from classical mechanics. The forces are defined in terms of potential-energy functions related to the internal coordinates of the molecule. The electrons of the molecule are not treated explicitly; rather, they provide a potential that is felt by the nuclei of the molecule. The forces collectively make up what is called the force field of the molecule. The *force field* is a potential-energy surface that expresses the energy of the molecule as a function of the atomic coordinates. The force field can be used to calculate the energies of different conformations of the molecule. A molecule at rest in its force field will adopt the lowest energy conformation that is accessible. The energy of any conformation can be calculated from a knowledge of the force field and the coordinates of the atomic constituents of the molecule. The energy thus calculated is often called the *strain energy:*

$$E_{strain} = E_{bond} + E_{angle} + E_{torsional} + E_{nonbonded} \qquad (13.1)$$

It is the sum of contributions from terms arising from bond-length stretching or compressing, bond-angle bending, torsional-angle twisting, nonbonded interactions, and possibly other factors. Other sources of strain energy that have been used include Coulombic interactions, hydrogen bonding, and dipole–dipole interactions. Each interaction (e.g., bond length) has an optimum value. Each term in the summation can be parameterized in many different ways with different functional forms and parameter values, and many chemists have published their ideas in this area. Force fields have been developed in order to reproduce structural geometry, relative conformational energies, heats of formation, crystal packing arrangements, and other properties. Each term in the equation is actually a summation over all occurrences of that particular interaction in the molecule.

In practice, the strain energy of a molecule is minimized by a numerical method such as *gradient search,* whereby the atomic coordinates are altered and the energy recalculated repeatedly. The objective of the computation is to find the conformation with the least strain energy spread throughout the structure. Force-field methods provide a rigid structure at rest in an energy minimum.

The broadest objective of molecular mechanics is to obtain a simple but widely applicable force-field model backed up with experimental observations that can accurately generate the structures (and thermodynamic properties) of organic molecules. This objective has been fulfilled to a significant degree, but work continues in a search for further improvements in accuracy and generality.

13.2   IMPLEMENTATION OF MOLECULAR MECHANICS

The Force Field

A variety of functional forms and parameters have been used to construct force fields. Perhaps the most widely used force field has been developed by Allinger and co-workers (Burkert and Allinger 1982). This force field is implemented in software available from the Quantum Chemistry Program Exchange.

Each type of intramolecular interaction included in the strain-energy equation is represented by a functional form that includes a restoring force. Each term is formulated so that it tends to decrease in magnitude as the bond length or angle or other parameter tends toward its optimum value. The terms entering into the strain-energy calculation can be classified as follows according to the number of atoms involved:

1. Forces that involve two atoms such as bond-length deformation and nonbonded interactions
2. Forces that involve three atoms such as bond-angle deformations
3. Forces that involve four atoms such as torsional-angle deformations

It is seldom necessary to consider forces that involve more than four atoms.

Bond Stretching   The atoms that are joined by each bond are considered to be masses joined by a spring. Hooke's law is used to represent the energy necessary to stretch or compress a bond from its optimum length:

$$E_{bond} = \sum_{i=1}^{N_{bond}} k_i \, (l_i - l_i^0)^2 \tag{13.2}$$

where

$\quad N_{bond}$ = number of bonds in the molecule
$\qquad k_i$ = a proportionality constant (which is dependent on the bond type and atom identities)
$\qquad l_i$ = actual bond length of bond i
$\qquad l_i^0$ = optimum bond length (which is dependent on the bond type and atom identities)

The values for the two parameters k_i and l_i^0 are developed during the generation of the force field. The strain energy of Equation (13.2) is symmetric about l_i^0, giving equal energy penalties to stretched and compressed bonds. This energy term has the appearance of a quadratic function with its minimum at l_i^0. The approximation is best when the bond length is very near the optimum, and it is poorer as $(l_i - l_i^0)$ increases. That is, bond stretching and compression actually are anharmonic, and this can be included in the bond-length term by including a cubic term $(l_i - l_i^0)^3$. The total bond-length deformation energy contribution to the strain energy is taken to be the sum over all the bonds in the molecule.

Equation (13.2) assumes that l_i^0 depends only on the identities of the bonded atoms and the bond order. Thus, for example, a carbon–carbon single bond is assumed to have the identical ideal bond length regardless of its surroundings. Although this is a simplification of reality, it does work in molecular mechanics rather well.

Bond-Angle Bending   A Hooke's law formulation is also commonly used for the representation of bond-angle strain:

$$E_{\text{angle}} = \sum_{i<j}^{N_{\text{angle}}} k_{ij}(\theta_{ij} - \theta_{ij}^0)^2 \tag{13.3}$$

where

N_{angle} = number of bond angles present in the molecule
k_{ij} = proportionality constant (dependent on the bond types and identities of the three atoms involved)
θ_{ij} = actual bond angle involving bonds i and j
θ_{ij}^0 = the optimum bond angle (dependent on the bond types and identities of the three atoms involved

Just as with the bond stretching term, the parameters k_{ij} and θ_{ij}^0 are determined during the development of the force field. The total contribution to the strain energy by bond-angle deformation is taken as the sum over all the angles in the molecule.

The amount of energy necessary to alter a bond angle is less than that required to stretch a bond, so the force constant in Equation (13.2) is larger (by approximately an order of magnitude) than that of Equation (13.3). Therefore, a molecular structure is more likely to have strained bond angles than strained bond lengths when it is in a conformation of low total strain energy.

Hooke's law is quadratic, so the contributions to strain energy from an equal-increment deformation are larger as the bond angle moves further from the optimum value. A better fit between strain and deformation would be possible if the force constant were made a function of $\Delta\theta = |\theta_{ij} - \theta_{ij}^0|$. An alternative way to correct the terms, which has been employed in some force fields, is to include a cubic term in Equation (13.3): $\Delta\theta^3$.

Equation (13.3) assumes that θ_{ij}^0 depends only on the identities of the three bonded atoms and the two bond orders. An sp^3-hybridized carbon atom bonded to two other sp^3 carbons will always have the same ideal bond angle, regardless of surroundings. This simplification is made so that the force field does not become too complex or the computations too complicated.

Including independent terms for bond-length and bond-angle deformations neglects the obvious interplay between these aspects of molecular conformation. As a bond angle is changed, the two associated bond lengths will tend to change also. This source of strain will appear in the calculation of energies between atoms in a 1–3 relationship (atoms each bonded to a common atom) due to through-space interactions if such a term is included in the force field. Alternatively, a cross term that includes both bond-length and bond-angle deformations multiplied together can be used. One form for a term that would take such interplay into account is as follows:

$$E_{sb} = \sum_{i<j} k_{ij}^{sb} \, (l_i - l_i^0 + l_j - l_j^0)(\theta_{ij} - \theta_{ij}^0) \tag{13.4}$$

where E_{sb} denotes a stretch-bend term, k_{ij}^{sb} is a new stretch-bend parameter, and the remaining terms have the same meanings as in the previous equations.

Torsional-Angle Twisting   Torsional angles (also called *dihedral angles*) play in important role in the conformation of organic molecules. The torsional angle term in the force field represents the energy cost of rotation about single bonds. The interaction involves the three bonds that join together four atoms. Label the atoms A, B, C, and D (Fig. 13.1). The torsional angle is defined as the angle measured about the B–C axis from the A–B–C plane to the B–C–D plane. There are two possible conventions as to which direction of rotation is defined as a positive angle and which is negative, and both conventions have been used. As long as a particular force field is consistent, the convention employed is immaterial.

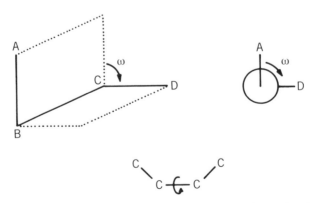

Figure 13.1   Conventions used in defining torsional angles for molecular modeling.

$$\mathbf{g}_i = \frac{\partial E_{\text{strain}}}{\partial x_i} \qquad (13.10)$$

$$\mathbf{G}_{ij} = \frac{\partial^2 E_{\text{strain}}}{\partial x_i \, \partial x_j} \qquad (13.11)$$

where the x_i and x_j represent the positions of the atoms in the molecule. The gradient vector and the Hessian matrix are evaluated numerically or analytically.

Steepest descent was the first method used by Wiberg (1965) and studied extensively by Burkert and Allinger (1982). This method depends on (1) either calculating or estimating the first derivative of the strain energy with respect to each coordinate of each atom and (2) moving the atoms. The derivative is estimated for each coordinate of each atom by incrementally moving the atom and storing the resultant strain-energy change. The atom is then returned to its original position, and the same calculation is repeated for the next atom. After all the atoms have been tested, their positions are all changed by a distance proportional to the derivative calculated in step 1. The entire cycle is then repeated. The calculation is terminated when the energy is reduced to an acceptable level. The main problem with the steepest-descent method is that of determining the appropriate step size for atom movement during the derivative estimation steps and the atom movement steps. The sizes of these increments determines the efficiency of minimization and the quality of the result.

The methods known as *parallel tangents* and *pattern search* are variants of the steepest-descent method. The parallel-tangents procedure starts by finding two new locations for an atom, each of which has lower strain energy than the initial point. Then a curve is fit through the three points and its minimum is found, and this fourth point is chosen for the new atom location. The cycle repeats. The pattern search method is similar to the steepest-descent method except that each atom is left at its best position as positions are tested. An additional advancement involves altering the step sizes taken by the atoms as dictated by the size of the derivative found during the search or by the estimated value of the second derivative.

An advantage of these first-derivative methods is the relative ease with which the force field can be changed. If the first derivatives are being estimated numerically, only the potential function itself appears in the programs. Therefore, it can be altered without ramifications elsewhere in the program.

The Newton–Raphson methods of energy minimization utilize the curvature of the strain-energy surface to locate minima. The computations are considerably more complex than the first-derivative methods, but they utilize the available information more fully and therefore converge more quickly. These methods involve setting up a system of simultaneous equations of size $(3n - 6) \, (3n - 6)$ and solving for the atomic positions that are the solution of the system. Large matrices must be inverted as part of this approach. The

Newton–Raphson methods are discussed in detail by Burkert and Allinger (1982).

A completely different method for seeking low strain conformations involves rotation about single bonds. In this approach, certain key single bonds are chosen for investigation. The torsional angles with these single bonds at their centers are rotated incrementally, with recalculation of molecular strain at each incremental torsional angle value. This methodology searches only a small portion of the conformational energy space, but it searches that portion more intensively than alternative energy minimization schemes.

Molecular Mechanics Advantages and Drawbacks

Molecular mechanics is an extremely widely used method for generating molecular models for a multitude of purposes in chemistry. In this section we discuss some of the reasons for this widespread use, but we also cover the drawbacks and limitations of the method.

The first major reason for the popularity of molecular mechanics is its speed, which makes it computationally feasible for routine usage. The alternative methods for generating molecular geometries, such as ab initio calculations or semiempirical molecular-orbital calculations, consume much larger amounts of computer time, making them much more expensive to use. The economy of molecular mechanics makes studies of relatively large molecules such as dyes or pharmaceuticals feasible on a routine basis. Economy also makes it feasible to use molecular mechanics in teaching situations where more expensive alternatives might not be feasible. Molecular mechanics applications to quite large molecules such as polypeptides and proteins is an ongoing field of research.

Molecular mechanics is relatively simple to understand compared to alternative methods. The total strain energy is broken down into chemically meaningful components that correspond to an easily visualized picture of molecular structure. Molecular mechanics calculations can easily be used to enhance classroom explanations of organic chemistry concepts such as steric hindrance, torsional barriers to rotation, conformations of macrocyclic rings, deformations of bond angles in strained molecules, and similar structural ideas.

When further detailed studies of a molecular require the input of a starting geometry, molecular mechanics can be used to get an initial conformation economically. Such an initial conformation can then be passed to more sophisticated methods such as quantum-mechanical calculations for refinement. Overall, this two-step process is more economical than are the alternatives.

Molecular mechanics also has some drawbacks or limitations that must be discussed. The first potential pitfall to be aware of is that molecular mechanics routines will generate a conformation for which the strain energy is minimized. However, the minimum found during a calculation may not be the global minimum, because the potential-energy surface is ordinarily quite convoluted. It is relatively easy for the procedure to become trapped in a local energy minimum. There are schemes for minimizing the risk of such entrapment in local minima; for example, the calculation can be done a number of times starting from different initial geometries to see if the final geometry found remains the same. Alternatively, a search of the potential-energy surface on a coarse grid can be done using torsional-angle incremental changes, followed by a more careful investigation of only those regions of the surface where minima are likely to be found. None of these methods, however, guarantees that a molecular mechanics calculation will find the global minimum in the potential-energy surface.

Another drawback is that any molecular mechanics routine is applicable only to molecular structures that are similar to those used for its parameterization. Thus, it is not uncommon for a user of a molecular mechanics routine to find that the molecular structure of interest cannot be modeled by the routine, which may announce that certain necessary parameters are unavailable. A more dangerous outcome is that the routine may provide incorrect geometries without any warning because of its parameterization.

An obvious drawback of the molecular mechanics approach is that it cannot be used to study any molecular system where electronic effects are dominant. Thus, many important facets of chemistry (e.g., bond breaking) are beyond the capabilities of molecular mechanics. Here, quantum-mechanical approaches that explicitly account for the electrons in molecules must be used.

Graphical Display of Molecules

An important capability of computers is their ability to present complex information in pictorial form. In the context of molecular mechanics, this translates into the ability to display molecules graphically on videodisplay terminals. Presentation of molecules for visual analysis allows the chemist to view the model and judge its quality or seek insights based on the structure. The topic of graphical display of molecular structures is covered in detail in Chapter 18.

Commercial Molecular Modeling Software

A large number of molecular modeling software packages are now available in the marketplace. As part of the continuing series, *Reviews in Computational Chemistry,* Boyd (1993) presents up-to-date lists of software for molecular modeling. The list is changing at a reasonably rapid pace, but its composition in Volume 4, 1993 is an indication of the status of the field.

Some of the better known general purpose molecular modeling software packages intended for use with personal computers are as follows: Alchemy III (Tripos Associates), Personal CAChe (CAChe Scientific), CAMSEQ/M (Weintraub Software Associates), Chem3D/Plus (Cambridge Scientific Computing), ChemCad+ (C−Graph Software), Chem-X (Chemical Design),

HyperChem (Autodesk), MOBY (Springer-Verlag), and PCMODEL (Serena Software). These packages are designed for the user who is a chemist but not an expert in molecular modeling. Thus, chemists who are not dedicated to modeling and its variants as a full-time job can still do sophisticated molecular modeling without having to learn all the idiosyncrasies of the more capable software systems. These personal-computer (PC) systems are also very well suited to educational usage in support of organic chemistry and biochemistry classes and experiments.

Some of the better-known general-purpose molecular modeling software packages intended for use with workstations are as follows: AMBER (Peter Kollman, UCSF), CAChe WorkSystem (CAChe Scientific), Chem-X, GRO-MOS (Biomos B.V.), Insight/Discover (BIOSYM Technologies), MacroModel (Clark Still, Columbia), MM3 (Technical Utilization Corp.); QUANTA/CHARMm (Molecular Simulations), and SYBYL (Tripos Associates). The workstation packages have capabilities beyond those of the PC products, and they are developed for users who are investigating large molecules or are especially interested in calculations such as solvation, molecular dynamics, three-dimensional fitting together of molecules, and other advanced computational tasks.

13.3 MOLECULAR DYNAMICS

Molecules are dynamic, undergoing vibrations and rotations continually. The static picture of molecular structure provided by molecular mechanics therefore is not realistic. The picture provided by x-ray structure determination is also not completely accurate in that one observes a time-averaged structure by this technique. The branch of computational chemistry that deals with the dynamics of molecular structure is *molecular dynamics*. Molecular dynamics methods have been applied to many types of molecular systems, ranging from bulk simple liquids to solvated proteins. The field is rapidly changing because of advances in the theory and methodology for tackling these problems and also the ever-increasing availability of affordable computational facilities, including workstations and supercomputers.

The forces that are acting on the atoms of any molecule are given by the first derivatives of the potential function with respect to the atom positions. In Section 3.2, we discussed the nature of empirical potential functions in the context of molecular mechanics. The potential functions used for molecular dynamics are similar to the molecular mechanics potential function. Thus, the classical equations of motion for the molecule of interest can be integrated numerically to solve for the atomic positions as a function of time. The time intervals used for the integration must be short with respect to the natural frequencies of vibration. The time intervals for the individual calculational steps are therefore on the order of femtoseconds (fs) ($1 \text{ fs} = 1 \times 10^{-15} \text{ s}$). The overall timescale accessible to molecular dynamics calculations is on the order of picoseconds (ps) (1 ps = 1×10^{-12} s). Calculations have been done for as long as 1000 ps = 1 ns (1 nanosecond = 1×10^{-9} s), but this is a long, expensive calculation.

Let us consider a solute molecule (a peptide or a protein) in a solvent as our molecule of interest. The starting positions of the atoms in the solute molecule are obtained from a known x-ray structure, and the solvent molecules surrounding it are taken from a preequilibrated box of solvent molecules. Then the x-ray structure of the solute molecule and the solvent atoms are submitted to a minimization procedure to relieve any strains inherent in the starting positions of the atoms. Such strains are likely to be present because the x-ray structure is an average structure, and some of the atoms will be in unrealistic positions that would yield very high forces on them.

The next step is to assign velocities to all the atoms. These velocities are drawn from a low-temperature Maxwellian distribution. The system is then equilibrated by integrating the equations of motion while slowly raising the temperature and adjusting the density. The temperature is raised by increasing the velocities of all of atoms. There is a simple analytical function expressing the relationship between kinetic energies of the atoms and the temperature of the system.

$$T(t) = \frac{1}{(3N - n)k_B} \sum_{i=1}^{N} m_i |v_i| \qquad (13.12)$$

where

$$\begin{aligned} T(t) &= \text{temperature of the system at time } t \\ (3N - n) &= \text{number of degrees of freedom in the system} \\ v_i &= \text{velocity of atom i at time } t \\ k_B &= \text{Boltzmann constant} \\ m_i &= \text{mass of atom } i \\ N &= \text{number of atoms in the system} \end{aligned}$$

This process of raising the temperature of the system will cover a time interval of 10–50 ps.

The period of heating to the temperature of interest is followed by a period of equilibration with no temperature changes. The stabilization period will cover another time interval of 10–50 ps. The mean kinetic energy of the system is monitored, and when it remains constant, the system is ready for study. The protein is in an equilibrium state at the desired temperature.

The molecular dynamics experiment is then started. This consists of letting the molecular system run free for a period of time, saving all the information about the atomic positions, velocities, and other variables as a function of time. This (voluminous) set of data is called a *trajectory*. The length of time that can be saved during a trajectory is limited by the computer time available (and the speed of the computer being used and its charging algorithm). Lengths of time in the range of 25 ps to a several nanoseconds are currently accessible, but this is changing rapidly as supercomputers become more widely available.

Once a trajectory has been calculated, all the equilibrium and dynamic properties of the system can be calculated from it (in principle). Equilibrium properties are obtained by averaging over the property during the time of the trajectory. Plots of the atomic positions as a function of time schematically depict the degree to which molecules are moving during the trajectory. The root-mean-square (RMS) fluctuations of all of the atoms in a molecule can be plotted against time to summarize the aggregate degree of fluctuation for the entire structure. Such a plot can be compared directly to a similar plot obtained from x-ray experimental data. The methods of molecular dynamics are becoming an extremely important component of the study of protein structures as investigators attempt to find structural reasons for protein activity and function.

REFERENCES

Allinger, N. L., "MM2(87) Molecular Mechanics 1987 Force Field," Quantum Chemistry Program Exchange, Program Number MM2(87) (1987).

Altona, C., and D. H. Faber, "Empirical Force Field Calculations. A Tool in Structural Organic Chemistry," *Top. Curr. Chem.*, **45**, 1–38 (1974).

Bowen, J. P., and N. L. Allinger, "Molecular Mechanics: The Art and Science of Parameterization," in *Reviews in Computational Chemistry*, Vol. 2, VCH Publishers, New York, 1991.

Boyd, D. B., and K. B. Lipkowitz, "Molecular Mechanics, The Method and Its Underlying Philosophy," *J. Chem. Ed.*, **59**, 269–274 (1982).

Boyd, D. B., "Compendium of Software for Molecular Modeling," in *Reviews in Computational ChemistryEN*, Vol. 4, K. B. Lipkowitz and D. B. Boyd (eds.), VCH Publishers, New York, 1993.

Brooks, C. L., M. Karplus, and B. M. Pettitt, *Proteins: A Theoretical Perspective of Dynamics, Structure, and Thermodynamics*, Wiley, New York, 1988.

Burkert, U., and N. L. Allinger, *Molecular Mechanics*, American Chemical Society, Washington, DC, 1982.

Haile, J. M., *Molecular Dynamics Simulation. Elementary Methods*, Wiley-Interscience, New York, 1992.

Karplus, M., and J. A. McCammon, "The Dynamics of Proteins," *Sci. Am.*, **254**, 42–51 (1986).

Kollman, P., "Theory of Complex Molecular Interactions: Computer Graphics, Distance Geometry, Molecular Mechanics, and Quantum Mechanics," *Accts. Chem. Res.*, **18**, 105–111 (1985).

Kollman, P., "Molecular Modeling," *Ann. Rev. Phys. Chem.*, **38**, 303–316 (1987).

Lipkowitz, K. B., "Empirical Force-Field Method," in *Conformational Analysis of Cyclohexanes, Cyclohexadienes, and Related Hydroaromatic Compounds*, P. W. Rabideau (ed.), VCH Publishers, New York, 1989, pp. 301–319.

Rasmussen, K., *Lecture Notes in Chemistry*, Vol. 27: *Potential Energy Functions in Conformational Analysis*, Springer-Verlag, Berlin, 1985.

Venkataraghavan, R., and R. J. Feldmann (eds.), *Macromolecular Structure and Specificity: Computer-Assisted Modeling and Applications*, New York Academy of Sciences, New York, 1985.

Westheimer, F. H., "Calculation of the Magnitude of Steric Effects," in *Steric Effects in Organic Chemistry*, M. S. Newman (ed.), Wiley, New York, 1956, Chapter 12.

Wiberg, K. B., "A Scheme for Strain Energy Minimization. Application to the Cyclohexanes," *J. Am. Chem. Soc.*, **87**, 1070–1078 (1965).

14

PATTERN RECOGNITION

14.1 PATTERN RECOGNITION METHODS

Chemists are accustomed to using graphical methods for data interpretation. Typically, a set of measurements of one dependent variable is made for different values of one independent variable while the remaining independent variables are held constant. The results are plotted as a two-dimensional graph for viewing and analysis. If the variables are related through a simple relationship, then a curve (perhaps even a straight line) can be fit to the set of data. This operation constitutes a generalization and an interpretation. The set of data is characterized tersely by the coefficients of the fit equation. The equation can also be used for prediction; that is, values of the dependent variable can be calculated from a value of the independent variable and the equation. Of course, the predictions of the dependent variable are most likely to be reasonable if the equation is used for interpolation between observed values of the independent variables. However, extrapolation may also be warranted in some circumstances.

Situations arise in science, however, in which independent variables cannot be varied one at a time while the others are held constant. There may be a very large number of independent variables to sort through. The data may be characterized by membership in categories rather than having a quantitatively measurable property. For example, a sample of stone comes from quarry 1 or quarry 2, but it can never come from quarry 1.37. In many cases, this type of data can still be represented with variables that can be visualized as points, but no longer in a two-dimensional space. An object or observation characterized by n descriptors can be represented as a point in an n-dimensional space.

Some examples of chemical data that can be expressed in this notation are as follows: (1) air-pollution particulate samples can be characterized by the trace-metal concentrations found in the samples—the number of values for a sample would be the number of trace metal concentrations measured; or (2) trace-level organic acid concentrations in human body fluids could be used to represent individual samples.

Thus, the data encountered in chemical problems can often be represented by a vector as

$$\mathbf{X}_i = (x_{i1}, x_{i2}, \ldots, x_{in}) \tag{14.1}$$

where each individual component of the pattern vector x_j is the value for the jth descriptor. To reference the ith observation of the data set, the notation $\mathbf{X}_i$ is used. An equivalent way to view this representation of data is to consider each observation to be represented by a point in an n-dimensional space.

For a given observation, which is represented by a given point, the value of each coordinate is just the numerical value for one of the descriptors comprising the representation. The expectation is that the points representing objects of a common category will cluster in one limited region of the space separate from those belonging to a different category. The clusters are regions of high local density relatively far apart from each other. *Pattern recognition* consists of a set of methods for investigating data represented in this manner to assess the degree of clustering and general structure of the data space.

Two classes of pattern recognition methods are supervised and unsupervised. *Supervised* methods employ a training set of patterns belonging to known classes to develop discriminants that separate the classes from each other. These discriminants are then available for predicting the classes of unknown patterns. *Unsupervised* methods do not have access to the class information but only to the patterns themselves. The goal is to investigate the pattern space and the arrangements of the points, by looking for clusters of pattern points that may be indications of meaningful relations among the clustered points.

In order to discuss clustering of points in a space, one must start with a definition of similarity. It is most common to use Euclidean distance in the definition of similarity in pattern recognition work. The Euclidean distance between the two patterns

$$\mathbf{X}_i = (x_{i1}, x_{i2}, \ldots, x_{in}) \tag{14.2}$$

and

$$\mathbf{X}_j = (x_{j1}, x_{j2}, \ldots, x_{jn}) \tag{14.3}$$

is given by the expression

$$d_{ij} = \sum_{k=1}^{n} (x_{ik} - x_{jk})^2 \tag{14.4}$$

The similarity between the two patterns is inversely related to the distance between them. Other inverse relations have also been used in different types of pattern recognition work, but a very common form is

$$S_{ij} = 1 - \frac{d_{ij}}{d_{max}} \tag{14.5}$$

where d_{max} is a normalizing factor. If d_{max} is the largest interpoint distance in the data set, S_{ij} spans the range from unity for identical patterns to zero for the least similar pair of patterns.

Example Data Set

The most effective way to introduce many of the concepts of pattern recognition will be to use a specific set of data for illustration. Therefore, a set of 100 patterns each consisting of five descriptors is presented in Figure 14.1. The first column refers to the class membership of each pattern, with 1 denoting the positive class and -1 denoting the negative class. There are 50 patterns in each class. The values of the raw descriptors range from 1 to 20. Statistics relating to the descriptors are found in Table 14.1, part A. These statistics refer to the overall set of 100 patterns, without regard to class membership. The simple correlations between each pair of descriptors are shown in Table 14.1, part B. The small values that appear in the correlation matrix show that the descriptors are almost completely orthogonal to one another, that is, these five descriptors are nearly independent. Figure 14.2 depicts this visually. It is a plot of descriptor 1 versus descriptor 2, and it shows a scattered pattern of values. The members of class 1 are shown as solid circles, and the members of class 2 are shown as open circles.

These 100 patterns belong to two classes. The means and standard deviations for each class are shown in Table 14.1, part A. An interesting test is to see how well the patterns can be put into their proper categories based on one descriptor at a time. The results of such a test are shown in Table 14.1, part C. Thresholds were inserted into the descriptors to maximize the number of correct classifications without regard to class. The best results obtained are for descriptor 1, with 30 of the 100 patterns classified incorrectly, which is a 70% correct classification rate. The classification rates for the other four descriptors are even lower than this. Thus, there is no one descriptor that does a very good job of separating these 100 patterns into their two classes. This set of 100 patterns will be used in the following pages to illustrate some of the pattern recognition methods discussed.

1	9.446	4.652	5.089	6.715	7.659
1	2.275	7.488	5.846	3.161	6.999
1	8.429	8.758	6.035	4.489	3.155
1	4.399	6.073	7.395	1.627	2.080
1	6.291	8.199	1.430	1.651	5.536
1	6.674	1.946	5.121	6.372	1.194
1	1.749	4.944	8.628	2.311	5.513
1	3.048	6.231	5.220	9.610	3.046
1	2.908	4.805	3.870	5.941	5.363
1	6.922	4.930	1.826	2.480	9.829
1	7.476	3.589	9.231	2.259	5.733
1	5.756	9.958	7.919	7.227	1.535
1	4.175	2.696	3.936	3.557	5.674
1	2.871	9.425	4.766	7.312	9.847
1	8.577	5.058	1.047	7.018	3.870
1	4.348	6.583	5.728	2.982	2.992
1	6.689	8.739	4.492	8.297	3.041
1	9.455	7.800	2.158	2.379	9.298
1	9.266	9.077	2.034	7.627	6.522
1	5.503	4.550	6.226	4.038	8.867
1	7.345	1.293	3.187	3.292	2.449
1	2.352	8.532	1.901	8.235	8.630
1	3.109	7.563	3.125	2.462	8.009
1	4.075	9.100	4.393	7.727	5.511
1	2.555	4.735	5.885	5.698	7.299
1	6.464	3.300	6.113	6.721	3.521
1	1.567	9.808	5.330	9.532	3.487
1	8.783	5.136	6.777	1.782	9.780
1	6.609	4.111	4.837	9.836	3.968
1	6.879	5.609	3.993	3.646	3.119
1	9.439	7.733	6.781	4.484	9.830
1	1.654	7.676	7.608	3.614	2.206
1	3.734	5.677	5.274	7.038	3.660
1	2.135	9.473	8.350	9.809	5.401
1	6.344	3.960	4.158	3.321	5.287
1	2.006	8.591	9.868	7.118	9.357
1	1.592	9.357	9.688	3.748	8.289
1	3.594	9.431	9.278	8.999	1.018
1	9.445	4.268	3.223	8.712	7.521
1	2.260	4.152	6.866	1.817	9.250
1	7.725	6.020	4.718	8.456	1.520
1	2.125	2.623	2.794	1.176	2.070
1	7.270	8.248	1.466	5.155	6.818
1	2.828	1.552	7.244	5.567	4.739
1	5.712	8.347	3.220	9.281	1.201
1	5.004	9.805	4.983	9.737	8.059
1	9.608	3.267	7.283	9.118	6.363
1	2.467	5.562	4.090	3.670	2.879
1	6.401	1.149	8.228	1.455	4.615
1	4.134	2.049	8.256	2.448	7.241
-1	1.047	7.768	1.232	9.566	9.657
-1	1.008	5.689	2.309	9.151	5.504
-1	8.184	2.320	1.038	9.559	3.555
-1	5.382	1.254	7.743	9.787	3.955
-1	2.261	4.320	4.996	7.056	9.340
-1	3.579	2.974	1.185	7.190	6.958
-1	4.085	1.623	3.737	7.432	5.965
-1	2.146	4.247	1.190	4.719	9.541
-1	1.177	1.362	1.915	8.731	6.713
-1	1.157	4.540	4.673	7.268	6.750
-1	2.749	1.998	1.341	5.190	8.152
-1	8.958	2.086	2.452	9.245	5.935
-1	2.093	1.533	9.708	8.097	7.790
-1	3.330	3.770	1.455	9.212	8.542

Figure 14.1 One hundred patterns of five variables per pattern. The first column indicates the class membership.

-1	2.654	3.293	7.049	9.879	3.001
-1	3.726	2.921	4.516	9.968	4.082
-1	3.092	1.436	9.442	9.286	4.120
-1	4.378	1.066	2.362	7.397	5.232
-1	8.009	2.085	1.943	9.197	4.167
-1	1.101	3.529	1.657	5.419	4.061
-1	7.158	1.457	1.034	9.176	9.725
-1	1.317	1.700	2.500	9.372	5.807
-1	2.056	4.475	1.524	6.913	7.857
-1	3.025	1.310	6.264	6.598	5.429
-1	1.737	8.103	2.569	9.755	9.521
-1	3.501	2.109	3.246	9.058	3.312
-1	2.221	4.699	3.316	7.243	6.017
-1	2.279	1.586	9.509	9.885	2.946
-1	2.623	1.391	4.690	8.046	8.887
-1	5.128	2.969	1.016	9.746	8.041
-1	2.153	2.851	8.124	9.884	5.044
-1	4.951	3.286	4.621	9.574	3.964
-1	3.533	1.038	9.379	8.235	5.738
-1	1.274	2.449	3.598	7.628	2.649
-1	2.007	3.298	2.301	9.787	1.103
-1	5.083	1.162	3.390	8.351	9.660
-1	2.302	2.236	1.904	8.819	3.910
-1	2.200	1.134	2.263	3.299	7.722
-1	1.737	1.964	2.830	5.420	8.458
-1	1.202	4.833	2.748	9.669	6.601
-1	3.478	3.282	1.898	5.178	8.770
-1	6.294	1.508	4.888	9.868	1.885
-1	1.394	1.421	2.657	6.356	6.314
-1	2.297	1.351	3.578	8.646	4.493
-1	4.130	2.838	8.205	9.652	6.082
-1	5.227	3.246	1.928	7.250	4.956
-1	4.458	1.216	6.749	8.130	5.664
-1	4.399	2.167	2.067	5.227	9.735
-1	4.763	1.766	5.146	9.342	9.912
-1	1.562	3.538	3.987	6.758	4.492

Figure 14.1.   (*Continued*)

The methods and techniques of pattern recognition can be subdivided into the following categories for simplification: preprocessing, mapping and display, classification by discriminants, clustering methods, and modeling methods. Each of these will be introduced in the following pages.

Preprocessing

Preprocessing operations alter the form of the representations of the observations and include operations such as scaling, normalization, and other mathematical transformations. The objective of all preprocessing is to facilitate the analysis phase of the pattern recognition study.

Scaling and normalization are required to convert the given units of measurements for different descriptors to a compatible form. For example, if one descriptor had a natural range of 1–10, and another had a natural range of 0.003–0.040, simultaneous analysis of these descriptors without scaling could give unwarranted and unwanted emphasis to one or the other descriptor. In

TABLE 14.1   Statistics for the Five-Dimensional Data Set

A.   Individual Descriptor Statistics

	Overall		"Yes" Class		"No" Class	
Descriptor	Mean	Standard Deviation	Mean	Standard Deviation	Mean	Standard Deviation
1	4.27	2.50	5.23	2.62	3.31	1.97
2	4.40	2.73	6.07	2.61	2.72	1.60
3	4.57	2.56	5.34	2.35	3.80	2.56
4	6.77	2.67	5.41	2.79	8.12	1.69
5	5.79	2.57	5.42	2.74	6.15	2.36

B.   Correlation Matrix

	1	2	3	4	5
1	1.0	0.067	0.057	0.108	0.048
2		1.0	0.087	0.108	0.022
3			1.0	0.082	0.182
4				1.0	0.136
5					1.0

C.   Classification by Individual Descriptors

	Number Incorrect		
Overall	"Yes" Class	"No" Class	Percentage
30	25	5	70
32	12	20	68
42	26	16	58
38	25	13	62
45	16	29	55

D.   Results of Eigenanalysis[a]

	Eigenvalue	Cumulative Variance (%)
1	1.208	24.4
2	1.182	48.3
3	1.002	68.5
4	0.883	86.4
5	0.675	100.0

[a]*Eigenvector 1:* $(-0.40004, -0.53842, -0.14597, 0.67739, -0.26438)$. *Eigenvector 2:* $(0.00252, 0.21052, 0.70373, 0.05655, -0.67620)$.

TABLE 14.1.   (*Continued*)

E.   Transformation of Pattern 1 and 39

Pattern 1[a]		Pattern 39[b]	
Descriptor Raw Value	Autoscaled Value	Descriptor Raw Value	Autoscaled Value
9.446	2.0715	9.445	2.0711
4.652	0.09289	4.268	−0.047656
5.089	0.20351	3.223	−0.52512
6.715	−0.020294	8.712	0.72794
7.659	0.72931	7.521	0.67559

[a]*Pattern 1:* Product of autoscaled values with eigenvector 1 = −1.115; product of autoscaled values with eigenvector 2 = −0.326.
[b]*Pattern 39:* Product of autoscaled values with eigenvector 1 = −0.41158; product of autoscaled values with eigenvector 2 = 0.79030.

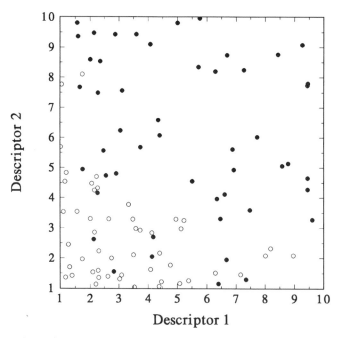

Figure 14.2   Plot of the values for descriptor 1 versus descriptor 2 for the test data set. Class 1 points are shown as solid circles, and class 2 points are shown as open circles.

addition, numerical instabilities in some algorithms can result from such imbalance among descriptor value ranges.

One preprocessing method that is widely used to correct such imbalances is called *autoscaling*. This method simultaneously scales and normalizes the data by translation so that for each descriptor, the mean becomes zero and the standard deviation becomes unity. The equation that describes autoscaling is

$$x'_{ij} = \frac{x_{ij} - \bar{x}_j}{\sigma_j} \tag{14.6}$$

$\bar{x}_j$ is the mean for the jth descriptor over the data set and σ_j is the standard deviation for the jth descriptor over the data set. The x_{ij} are the original, raw data points, and the x'_{ij} are the autoscaled data.

For descriptor 1 in the example data set, the mean value is 4.27 and the standard deviation is 2.50. Thus, the raw value of 9.45 for the first pattern would autoscale to $(9.45 - 4.27)/2.50 = 2.07$.

Mapping and Display

One powerful way of analyzing the structure of a data set is to map the points from the high-dimensional space to a two-dimensional space (a plane) for direct viewing. It has long been recognized that people are adept at recognizing patterns in two-dimensional plots, and this has been exploited in many pattern recognition display methodologies. A number of useful and imaginative graphical techniques exist for directly displaying multivariate data in just one or two dimensions. Their use has been neglected in the physical sciences in favor of more common rectangular plots and histograms. However, display techniques employing metroglyphs, linear and circular data profiles, Andrews plots, and Chernoff faces can be valuable tools for visualizing relationships between observations, for identifying outliers, and for classification purposes (e.g., Wang 1978).

A method that combines both mapping and display operations along with feature selection is principal-components analysis, also known as *Karhunen–Loeve transformations* (Tou and Gonzalez 1974). This transformation involves performing an eigenanalysis on the variance–covariance matrix of the data set, and then using the eigenvectors to perform a linear rotation. The multidimensional data set can be plotted in the two dimensions determined by the two principal eigenvectors and displayed for visual analysis. Alternatively, the data can be rotated by multiplication with any desired number of eigenvectors to reduce the dimensionality of the data set. This operation is commonly done using just enough of the available eigenvectors to retain 95 or 99% of the total variance of the data set. An advantage of this feature selection procedure is that the number of descriptors per observation is reduced, thus making further analysis more convenient in some ways. A disadvantage is that the individual descriptors resulting from this transformation are mixtures of the original descriptors, thus complicating analysis of later results.

When the example data set was submitted to a principal-components analysis, the results shown in Table 14.1, part D, were obtained. The eigenvalues were sorted by value from largest to smallest. The cumulative variance column shows that only 48% of the overall variance of this set of data is represented in the first two principal components. The almost equal size of the eigenvalues shows that this set of five-dimensional data nearly spans the space completely. The first two eigenvectors are shown. Each has been normalized to unit length. Vector multiplication of these two eigenvectors by the five-dimensional (5D) pattern vector for individual patterns of the example data set yields two-dimensional (2D) points, as shown in Table 14.1, part E. This is the Karhunen–Loeve transformation from 5D space to 2D space for individual patterns. When this operation is done for all 100 patterns, the results are as shown in Figure 14.3. This is called a *principal-components plot*. The x axis of the plot represents principal component 1, which is the direction in the 5D space that contains the most variance. The relative percentage of the total variance explained by the first principal component is 24.4% as shown in Table 14.1, part D. The y axis of the plot, principal component 2, is that direction orthogonal to principal component 1 that contains the most addi-

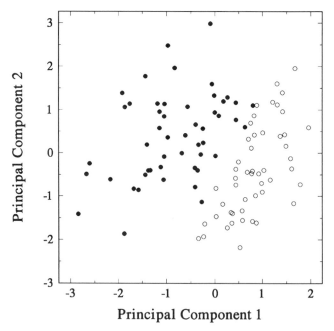

Figure 14.3 Plot of the 100 points of the test data set defined by their first two principal components. Class 1 points are shown as solid circles, and class 2 points are shown as open circles.

tional variance of the data set, namely 23.9%. Even though the data were not separable into their two classes to an appreciable degree based on any one descriptor, the principal-components plot of Figure 14.3 shows a great deal of separation. Thus, the Karhunen–Loeve transform has mapped a 5D space that could not be viewed directly to a 2D space that could be viewed.

Another mapping strategy is to attempt to find a 2D display where each point in the original data set is represented by a point in the 2D plane. This involves attempting to represent a high-dimensional space directly in two dimensions. Of course, a certain amount of error in the mapping is to be expected because an exact mapping is not possible. Many papers in the literature report approaches to the nonlinear mapping problem. Iterative nonlinear mapping routines have been implemented using the error function

$$E = \frac{\Sigma(d_{ij} - d'_{ij})^2}{\Sigma d'_{ij}} \tag{14.7}$$

where d'_{ij} is the Euclidean distance between points i and j in the original n-dimensional space and d_{ij} is the two-dimensional interpoint distance. The steepest-descent method, for example, can be used to minimize E as a function of the new interpoint distance. In a perfect mapping, where the n-space (n-dimensional space) is perfectly represented in 2D space, E would be zero. Nonlinear mapping algorithms are notorious for consuming large amounts of computer time. The starting estimates of the 2D-space point locations must be entered into the algorithm by the user. Both the final results obtained and the computer time invested in attaining the results are strong functions of the quality of the initial estimates entered. They might be the results of a previous principal-components analysis, for example. The results of the nonlinear mapping algorithm are usually presented graphically for visual examination.

Classification by Discriminants

It is very common for data sets being studied by pattern recognition methods to be category data. That is, each observation is tagged by its membership in a discrete category (e.g., quarry 1 or quarry 2). Then, pattern recognition methods are used to examine the points in the high-dimensional space to see if the point sets are disjoint. In other words, the goal is to discover whether the points belonging to one category are in a cluster separated from the points belonging to another cluster. One way to ascertain whether this structure is present involves putting discriminant surfaces through the space to slice it into regions. The search for powerful discriminants that indeed classify the points correctly constitutes the operation of classification and discriminant development.

A discriminant function can also be thought of as a decision surface passing through the n-space that contains the points of the data set. If the surface is flat (i.e., if it is a hyperplane), it can be completely and unambiguously represented by a normal vector perpendicular to it. With no loss of generality, the decision surface can be constructed to pass through the origin of the space. Then, every vector from the origin defines a plane that is the locus of points perpendicular to the normal vector. We will call the vector representation of a decision surface the *weight vector.*

Since it is very convenient to have the decision surface pass through the origin of the space, it is worthwhile to ensure that this will always be possible. An extra, orthogonal dimension is added to the pattern space, and all the pattern vectors are augmented with a $(d + 1)$st component. The $(d + 1)$st component of the pattern vectors can be assigned any value, but it is usually given the value of unity.

In addition to being able to represent a linear decision surface by specifying its normal vector, another important feature arises for this simple situation. The dot product of the normal vector $\mathbf{W}$ and a pattern vector $\mathbf{X}$ defines on which side of the hyperplane a given pattern point lies:

$$\mathbf{W} \, \mathbf{X} = |\mathbf{W}| \, |\mathbf{X}| \cos \theta \qquad (14.8)$$

where θ is the angle between the two vectors.

$$\begin{aligned} \cos \theta > 0 \ \text{ for } \ -90° < \theta < 90° \\ \cos \theta < 0 \ \text{ for } \ \ \ 90° < \theta < 270° \end{aligned} \qquad (14.9)$$

Since the normal vector is perpendicular to the hyperplane decision surface, all patterns having dot products that are positive lie on the same side of the plane as the normal vector, and all those with negative dot products lie on the opposite side. Points with zero dot products lie in the plane, which constitutes another definition of the location of the plane. (In a computational sense, the probability of a point lying within the plane is very small, with the exact probability depending on the number of significant digits used in the floating-point system of the computer being employed.) A second and equivalent way to define the dot product of two vectors is

$$\mathbf{W} \, \mathbf{X} = |\mathbf{W}| \, |\mathbf{X}| \cos \theta = w_1 x_1 + w_2 x_2 + \cdots + w_d x_d + w_{d+1} \qquad (14.10)$$

Each of the components of the weight vector $\mathbf{W}$ weights each of the terms of $\mathbf{X}$. This pairwise product version of the dot product is used in the actual computations of dot products in the software routines that implement the methods being described.

Although decision surfaces need not be linear, their simplicity when linear is appealing. Additionally, it can be shown that more complex decision surfaces can be implemented by linear decision surfaces preceded by appropriate preprocessing. That is, the space can be warped just as easily as the decision surfaces.

Two different approaches to the development of discriminants have been developed: parametric and nonparametric. *Parametric* methods of pattern recognition attempt to find classification surfaces or clustering definitions based on the statistical properties of the members of one or both classes of points. For example, Bayesian discriminants are developed using the mean vectors for the members of the classes and the covariance matrices for the classes. If the statistical properties of the classes cannot be calculated or estimated, then nonparametric methods can be used. *Nonparametric* methods attempt to find clustering definitions or classification surfaces by using the data themselves directly, without computing statistical properties. Examples of nonparametric methods include error-correction feedback linear learning machines (perceptrons), simplex optimization methods of searching for separating classification surfaces, and the k-nearest-neighbor method.

Parametric Discriminant Development   The parametric pattern recognition methods use the mean vectors and covariance matrices (or other statistical measures) of the two classes into which the patterns fall as their basis for development of discriminants (Tou and Gonzalez 1974). One parametric approach implements a quadratic discriminant function using the Bayes theorem. The equation expressing these discriminants is

$$d_k = \ln p_k - \ln \mathbf{C}_k - (\mathbf{X} - \mathbf{m}_k)^{\mathrm{T}} \mathbf{C}_k^{-1} (\mathbf{X} - \mathbf{m}_k) \quad k = 1,2 \qquad (14.11)$$

where p_k is the a priori probability for class k, $\mathbf{C}_k$ is the covariance matrix for class k, and $\mathbf{m}_k$ is the mean vector for class k. This discriminant assumes a multivariate normal distribution of the data. A pattern $\mathbf{X}_i$ is placed in the class for which $d_k(\mathbf{X})$ is greatest. Application of the Bayes quadratic discrimination method to the example data set yields a classification success rate of 99%, that is, only one pattern is classified incorrectly out of the 100.

If the assumption is made that the covariance matrices for the two classes are the same, the discriminant function developed using the Bayes theorem and the multivariate normal assumption simplifies from that above to

$$s = d_1(\mathbf{X}) - d_2(\mathbf{X}) = \ln p_1 - \ln p_2$$
$$+ \mathbf{X}'\mathbf{X}^{-1} (\mathbf{m}_1 - \mathbf{m}_2) - \mathbf{m}_1' \, \mathbf{C}^{-1}\mathbf{m}_1 + \mathbf{m}_2' \, \mathbf{C}^{-1}\mathbf{m}_2 \qquad (14.12)$$

where $s > 0$ for one class and $s < 0$ for the other class. The actual numerical computation can be done in any one of the three ways: using the covariance matrix for the entire data set for $\mathbf{C}$, using the covariance matrix of class 1 for $\mathbf{C}$, or using the covariance matrix for class 2 for $\mathbf{C}$. These computations were all done with the example data set, and the success rates for classification were 96, 96, and 97%, respectively. A few more patterns were misclassified by these linear discriminants than by the quadratic one, but the success rate was still rather good.

Nonparametric Discriminant Development   The nonparametric pattern recognition programs develop their discriminants using the training set of patterns to be classified rather than statistical measures of their distributions.

The k-nearest-neighbor (KNN) method is conceptually the simplest. An unknown pattern is assigned to the class to which the majority of its nearest neighbors belong. The metric that is used to determine proximity is ordinarily the Euclidean metric, but any measure can be used. The classification decisions are based on the calculation of point-to-point distances.

The KNN method was applied to the example data set with the following results. Calculation of the distances from pattern 1 to each of the other 99 points is the first step in the classification of pattern 1. Then these distances are inspected, and it turns out that point 39 is nearest to point 1. The actual distance is computed as the square root of the sum of the squared differences between the autoscaled values of the descriptors (shown in Table 14.1, part E). The actual distance between points 1 and 39 is 1.055. The second closest point is number 47 at a distance of 1.435, the third closest is 31 at a distance of 1.767, and so on. The class of the first-nearest-neighbor point is class 1, so the first-nearest-neighbor method (1NN) would classify point 1 as being in class 1. The closest three points are all in class 1, so the 3NN method would also classify point 1 as being in class 1. When the procedure is applied to all 100 points, the results are as shown in Table 14.2. Approximately 90% of the patterns are correctly classified using the KNN method.

Another widely used nonparametric method is the linear learning machine or perceptron (Tou and Gonzalez 1974). The algorithm for developing the discriminant is heuristic. A decision surface is initialized either arbitrarily or using the result from another linear discriminant development routine. Then the learning machine classifies one member of the training set at a time. When the current discriminant function correctly classifies a pattern, the discriminant is left unchanged. However, whenever an incorrect classification is made, the discriminant is altered in such a way that the error just committed is eliminated. The learning machine continues to classify the members of the training set repeatedly until no errors are committed or until the routine is externally terminated. This algorithm for the development of a linear discrimi-

TABLE 14.2   Results from K-Nearest-Neighbor Classifications

	Total	Correct	Incorrect	Percentage
First-Nearest-Neighbor Classification (1NN)				
Class 1	50	44	6	88
Class 2	50	47	3	94
Total	100	91	9	91
Three-Nearest-Neighbor Classification (3NN)				
Class 1	50	42	8	84
Class 2	50	48	2	96
Total	100	90	10	90

nant function has the desirable property that if a training set of patterns is separable into the two classes; that is, if a solution exists, this method will find a solution. This existence proof enormously enhances the attractiveness of the linear learning machine.

The dot product of the weight vector and a pattern vector gives a scalar whose sign indicates on which side of the decision surface the pattern point lies:

$$\mathbf{W}\,\mathbf{X} = s \tag{14.13}$$

(An arbitrary decision must be made as to which subset of the data is to be called the positive class and which is to be called the negative class.) When pattern i of the training set is misclassified, we obtain

$$\mathbf{W}\,\mathbf{X}_i = s \tag{14.14}$$

in which s has the incorrect sign for classifying $\mathbf{X}_i$. The object is to calculate an improved weight vector $\mathbf{W}'$ such that

$$\mathbf{W}'\mathbf{X}_i = s' \tag{14.15}$$

where the sign of the scalar result s' is opposite what it was previously. The new weight vector is calculated from the old one by adding an appropriate multiple of s to it:

$$\mathbf{W}' = \mathbf{W} + c\mathbf{X}_i \tag{14.16}$$

Combining these equations gives

$$s' = \mathbf{W}'\mathbf{X}_i = (\mathbf{W} + c\,\mathbf{X}_i)\mathbf{X}_i \tag{14.17}$$

which can be algebraically rearranged to give

$$c = \frac{s' - s}{\mathbf{X}_i\,\mathbf{X}_i} \tag{14.18}$$

It remains only to choose a value for s' to complete the derivation. An effective method is to let $s' = -s$. This moves the decision surface so that after the feedback correction the point being classified, $\mathbf{X}_i$ is the same distance on the correct side of the new decision surface as it was previously on the incorrect side of the old decision surface. If $s' = -s$ is put into the equation, then

$$c = \frac{-2s}{\mathbf{X}_i \mathbf{X}_i} \tag{14.19}$$

and **W'** can be calculated directly using the equations.

The training procedure involves iterating over all the pattern points in the training set and correcting the weight vector whenever an error is committed until the discriminant function converges on one that correctly classifies all the points. The process is called "learning" because the decision surface improves its performance at the classification task as its experience increases.

As mentioned above, this error-correction procedure can be shown to find a solution if one exists. Therefore, the weight vector can be initialized arbitrarily, although it is obviously better practice to use whatever information is available to estimate a starting weight vector.

The classification results obtained with linear discriminants are strongly affected by the ratio of the training set size n to the number of variables per observation d. This point has been investigated in some detail and reported (Stouch and Jurs 1986). The probability of correctly classifying 100% of the members of a training set due to chance is low for $n/d > 3$, but substantial classification success above the random expectation of 50% can still be obtained. For example, when $n/d = 5$, the probability is 0.5 that 77% of the members of the training set will be correctly classified, as a result of chance alone. The results reported in recent papers place limits on the types of problems that can be attacked by this type of pattern recognition approach, and they provide measures by which classification results can be judged.

A linear learning machine program (program LM) is included at the end of this chapter along with the set of example data and the results obtained during development of a decision surface separating the data set into its two classes.

Another nonparametric method develops a linear discriminant function through an iterative least-squares approach. The following error function is minimized:

$$Q = \sum_{i=1}^{m} [Y_i - F(s_i)]^2 \tag{14.20}$$

where

m = total number of patterns in the training set
Y_i = +1 for one pattern class and Y_i = −1 for the other class
s_i = dot product of the weight vector (discriminant) with the ith pattern
$F(s_i)$ = hyperbolic tangent function of s_i

The hyperbolic tangent function has the value of $+1$ for all positive values of s_i and -1 for all negative values of s_i. Therefore, minimizing Q is equivalent to minimizing the number of incorrect classifications. The function is nonlinear in the independent variables and thus cannot be solved directly, so an iterative algorithm is employed. This routine has been found to be particularly useful for dealing with data that are not completely separable into the two classes.

The search for a good discriminant that separates two clusters of points in an n-dimensional space from one another can be formulated as a linear programming problem (Ritter et al. 1975). The objective function whose value is to be minimized can be defined as the fraction of the training set of patterns that are incorrectly classified. Because two different discriminants can easily have the same classification power, it has been suggested that a secondary objective function be defined as the sum of the distances from the discriminant plane to the misclassified points. The secondary objective function is only invoked if the primary function is equal for two different discriminants being compared.

Clustering Methods

A subset of pattern recognition methods consists of clustering techniques that attempt to determine structural characteristics of a set of data by organizing the data into subgroups, clusters, or hierarchies. These methods are unsupervised because they attempt to define clusters solely on the basis of criteria derived from the data themselves, without advance knowledge of the classes of the data. This situation is in contrast to methods called *supervised* since they do have advance knowledge of which data points belong to which categories.

For clustering purposes, a quantitative measure of similarity between the d-dimensional points forming the data set must be defined. Similarity can be measured by Euclidean distance, squared Euclidean distance, correlation coefficient, or other suitable measures. For an entire set of data, an $(n \times n)$ similarity matrix contains all the pairwise interpoint similarities. For real-data sets, the points that represent objects that are similar may not only be relatively close together but may also form clusters. A cluster is a region of space with a high local density of points. The structure of data represented in this manner can be studied with cluster analysis methods. The objective of clustering is to generate a set of classes given only the data themselves and a quantitative definition of similarity. In this approach to multivariate data analysis, the individual points are usually not labeled as to class. Rather, the objective is to discover the class memberships. Thus, clustering works by seeking groupings of points that form natural clusters by examining the interpoint similarities systematically. An excellent discussion of the most important clustering methods can be found in Jain and Dubes (1988).

Hierarchical clustering methods construct dendrograms that express the structure of the data set. To start, the two most similar objects are found by scanning the similarity matrix. Then, these two objects are combined to form a new, combined object (a small cluster). The similarity matrix is reduced by 1 in size. A decision must be made as to how to calculate the similarity between the cluster and the other points, and later between two cluster of points. Nearest neighbors between the two clusters can be used, or the average distance from one cluster to another, or the centroids of the clusters being compared. The agglomeration process is repeated until all the points are linked. The overall classification hierarchy can be depicted as a treelike structure called a *dendrogram*. The clustering results obtained are wholly dependent on the normalization or transformation of the data used, on the similarity measure used, and on the agglomeration strategy used. Thus, these methods are experimental and exploratory in nature, and are most commonly used to explore the structure of a data set interactively. Hierarchical clustering methods are widely available in commercial software packages.

Nonhierarchical methods generate clusters without developing a tree structure. Two well-known methods are the K-means and isodata methods, presented here as examples of nonhierarchical clustering routines. The K-means method starts with k initial cluster centers supplied from outside the algorithm. The data points are distributed among the k centers with each point considered to be a member of the nearest cluster center. After this initial assignment step the cluster centers are updated to be the centroid of the points forming each cluster. Then the points are redistributed among the centers, the centers are updated, and so on, until the partitioning is stable. The output is the identity of the data points forming each cluster, statistics for each cluster, and intercluster distances. K-means is a relatively simple algorithm with few adjustable parameters.

Isodata (iterative self-organizing data analysis technique A) is a relatively complicated algorithm with many adjustable parameters, including the desired final number of clusters. It starts with a set of initial cluster centers supplied from outside. Each point is assigned to the nearest cluster center, but clusters with fewer than the specified number of points are discarded. The cluster centers are then updated to the centroid of the members. Clusters can be lumped together if they are too close together, or if there are too many clusters. Clusters can be split apart, on the other hand, if they are too diffuse, or if there are too few clusters. The lumping and splitting operations are controlled by user-supplied adjustable parameters. The number of clusters desired is input as a parameter, and the minimum number of points that can support a cluster is also specified. The output of Isodata is the identity of the data points forming each cluster, statistics for each cluster, and inter-cluster distances.

A tutorial article covers clustering in chemistry (Senn 1989). Some recent examples of the uses for clustering in chemical problems include the following: chemical information systems (Willet 1987), selection of compounds for biological testing (Willet et al. 1986), classification of meteorites (Massart et al. 1982), composition of atmospheric particles (Shattuck et al. 1985), characterization of gas-chromatographic stationary phases (Huber and Reich 1984), and studies of IR spectra (Zupan 1982).

Modeling Methods

A pattern recognition method has been developed by Wold and colleagues called SIMCA (Wold 1976, Wold and Sjostrom 1977). The SIMCA method uses principal-components analysis to construct an individual principal-component model to describe each class within the data set. The model for each class uses as many principal components as are necessary to describe that class of the data set adequately. The observations are classified according to their orthogonal Euclidean distance from the principal-component class models, that is, according to the magnitude of their residuals. In a multivariate classification problem, a new observation is classified as a member of the nearest model, and if it is distant from all the class models, then it is classified as an outlier. Thus, SIMCA can warn the user by labelling a new observation as an outlier or as a member of a new class not considered during the construction of the class models. SIMCA does not calculate discriminants; rather, it generates individual models that describe the membership of each class of points and classifies new points by proximity to these class models. SIMCA has been used in a large number of chemical investigations in recent years.

Program LM

Program LM and the subroutines TRAIN and PRED implement the error correction feedback linear learning machine algorithm for development of a linear binary pattern classifier. The mathematics involved in the algorithm has been presented in the section on nonparametric discriminant development. The main program, LM, inputs a set of data, randomly chooses a training set, initializes a number of parameters, calls the training routine, and calls the prediction routine.

The program presented here is similar to a program previously published (Jurs and Isenhour 1975), but this version has been changed in some ways. It has been made fully interactive, with the output all being sent to device 1. The data set to be analyzed is taken from file 5. All other parameters are initialized inside the routine for simplicity. Of course, all the parameters could be input by the user on request through a series of READ statements if desired.

The array named DATA contains the raw data or patterns. It is dimensioned for up to 100 patterns of up to five descriptors per pattern. The data are input from an input file named INPUT. No normalization of the data is performed by LM. The variable named LIST contains the category of each pattern, with $+1$ for one class and -1 for the other class. This information is input along with the data themselves. A training set of 80 patterns is chosen randomly, and the remaining 20 patterns are put in the prediction set. The weight vector is initialized so that each component has a value of zero except for the $(d+1)$st component with a value of unity. NTRSET is the number of patterns forming the training set, here 80; NUM is the number of descriptors per pattern, here 5. NPASS contains the number of error correction feedbacks that will be allowed before the learning routine is terminated because of lack of convergence, here 1000. TSHD is the nonzero threshold that is to be used during training, here 0.75. This has the effect of giving the decision surface being developed a thickness. NCONV is a flag variable that will report back to the calling program whether convergence was attained. IDTR is an array that contains the pattern numbers of the patterns forming the training set.

Subroutine TRAIN implements the error-correction training procedure for seeking a binary pattern classifier. To be correctly classified, a point must not only be on the proper side of the decision surface but also outside the actual volume occupied by the decision surface. The subroutine uses a subsetting procedure to attempt to make progress in the overall task as efficiently as possible. On the first pass through the data set, the entire training set is used with feedback corrections being made as necessary. At the same time, the routine keeps track of which members of the training were set misclassified. This information is stored in variable NSS. On the second pass, only those patterns missed on the first pass are classified, and a third subset is constructed. The sequence repeats until no patterns are misclassified, that is, until the subset is empty. Then the entire training set is classified, and the entire sequence begins again. The training program prints out the number of patterns misclassified on subsequent passes through the subsetting procedure. Only when two zeros appear in sequence is classification of the training set perfect. Subroutine TRAIN terminates when all the members of the training set are classified correctly or the allowed number of feedbacks is exceeded. Then it prints out the weight vector and the number of feedbacks employed during the training.

Subroutine PRED uses the trained weight vector to predict the classes of the members of the prediction set. Two calls to PRED are included in LM, one with the threshold value of 0.75 and one with the threshold value set to zero. PRED prints out the results of the predictions.

Program LM was executed with the example data set, and the output is presented. The identities of the 20 patterns chosen for the prediction set are output. The training routine outputs the number of patterns misclassified during each pass through either the entire training set or the subsets as a means to monitor progress during training. When two zeros appear, this means that training is complete. Then the weight vector is output, and the total number of feedbacks is printed. The prediction routine is called twice. With the threshold set to 0.75, 3 of the 20 patterns in the prediction set were not predicted because they were found to fall within the volume occupied by the decision surface. All the remaining 17 patterns were classified correctly. With the threshold set to zero, all 20 were correctly classified.

14.2 SELECTED CHEMICAL APPLICATIONS OF PATTERN RECOGNITION

Application studies of chemical problems using pattern recognition techniques have been reported in a number of areas. An early book by Jurs and Isenhour (1975) contains 113 references in all, of which at least 60 refer to chemical problems, and the review article by Kryger (1981) contains 130 references. The book by Varmuza (1980) contains 435 references. The progress of the field can be traced with the aid of a series of seven review articles published at 2-year intervals in *Analytical Chemistry*. These articles cover noteworthy advances within each 2-year interval: Kowalski (1980), Frank and Kowalski (1982), Delaney (1984), Ramos et al. (1986), Brown et al. (1988), and Brown et al. (1990, 1992). Detailed descriptions of the methods used in multivariate analysis in analytical chemistry can be obtained from a the following two textbooks: Sharaf et al. (1986) and Massart et al. (1987). Thus, there is a wealth of primary literature dealing with chemical applications of pattern recognition. This section will summarize some of the areas of application and provide references to a sampling of the primary publications. The best way for an interested reader to sample the information available in this area would be to consult one of the reviews.

Spectral Data Analysis

The elucidation of chemical structure information from spectral data is a long-standing problem of chemistry. This is the area first studied and most intensively studied using pattern recognition. Studies have been done with mass spectra, infrared spectra, Raman spectra, electrochemical data, γ-ray spectra, proton and ^{13}C NMR spectra, and Auger spectra.

Mass Spectrometry The analysis of low-resolution mass spectral data to obtain structural information was the first chemical problem studied by pattern recognition methods (Raznikov and Talroze 1966). Papers dealing with mass-spectrometric analysis are so numerous that Varmuza (1980) listed 14 review papers on the topic. The majority of the work has dealt with the recognition of molecular substructures or functional groups from the mass spectra of compounds. Some example studies are as follows: sequence analysis of oligodeoxyribonucleotides (Burgard et al. 1977), interpretation of steroid mass spectra (Rotter and Varmuza 1978), and determination of molecular structure parameters (Jurs et al. 1970).

Infrared Spectra The identification of function groups from the infrared spectra of organic compounds has been reported (e.g., Woodruff and Munk 1977). The infrared spectra are ordinarily digitized into equally spaced subintervals in order to represent the spectra as pattern vectors. Binary-coded spectra with pattern vector components restricted to 1 or 0 have also been studied.

NMR Spectra   Both proton and ^{13}C NMR spectra have been studied, and one review has appeared (Wilkins and Jurs 1978). Example studies include the influence of substituent effects on NMR patterns (Edlund and Wold 1980) and separation of structural classes (Brunner et al. 1975).

Classification of Complex Mixtures

Materials or mixtures can be classified into categories (e.g., origin) by pattern recognition. Examples from a number of diverse areas can be found in the literature: manufacturers and grades of paper (Duewer and Kowalski 1975), quarry sites of archaeological artifacts (McGill and Kowalski 1977), sources of atmospheric particulate matter (Gaarenstroom et al. 1977), classification of wines (Kwan and Kowalski 1980), determination of olive-oil origin (Forina and Tiscornia 1982), identification of crude-oil samples (Clark and Jurs 1979), determination of the clinical status of patients from urine samples (Rhodes et al. 1981), classification of cancer cells (Jellum et al. 1981), prediction of aircraft engine malfunction (Hancock and Synovec 1989), classification of disease states of crabs (Gemperline et al. 1992), identification of archaeological materials (Hayek et al. 1990), and classification of honeybees (Lavine and Carlson 1990). Pattern recognition methods have been applied to flavor and food research (Aishima and Nakai 1991, de Jong 1991).

Prediction of Properties from Molecular Structure

A number of studies of the application of pattern recognition to the problem of searching for relationships between molecular structure and biological activity have been reported. A large fraction of this type of research is involved with the generation of appropriate descriptors from the molecular structures available. Areas of study that have been reported in the literature include drug structure–activity relationships (SAR), studies of chemical communicants and studies of structure–toxicity relationships. Early applications of pattern recognition to drug design have been reviewed by Kirschner and Kowalski (1979). This area of application of pattern recognition is briefly mentioned by Varmuza (1980). A book describing one approach to SAR research has appeared (Stuper et al. 1979).

A few representative SAR studies are as follows: a study of 200 drugs for anticancer activity (Kowalski and Bender 1974), a study of 9-anilinoacridines for antitumor selectivity (Henry et al. 1982), studies of drugs of accepted therapeutic value (Menon and Cammarata 1977), structure–carcinogenicity potential (Rose and Jurs 1982), olfactory quality of organic compounds (Jurs et al. 1981), and structure–carcinogenic potential of PAH (Norden et al. 1978). An overview of many applications is provided by Miyashita et al. (1993).

```
      PROGRAM LM
C....
C.... BASIC LEARNING MACHINE PROGRAM
C....
      DIMENSION DATA(5,100),W(6),LIST(100),IDTR(100),IDPR(100)
     X    ,ITEMP(100)
      COMMON /IOUNIT/ NINP,NOUT
      OPEN (5,FILE='INPUT')
      NINP = 5
      NOUT = 6
      WRITE (NOUT,1)
    1 FORMAT (' LEARNING MACHINE PROGRAM',/)
C.... INITIALIZE RANDOM NUMBER GENERATOR
      IX = 521
      RR = RAND(IX)
C.... SET VALUES OF PARAMETERS
      NTRSET = 80
      NPRSET = 20
      TSHD = 0.75
      NTOT = NTRSET + NPRSET
      NPASS = 1000
      NUM = 5
C.... INPUT DATA SET
      DO 10 I=1,NTOT
   10 READ (NINP,11) LIST(I),(DATA(J,I),J=1,NUM)
   11 FORMAT (I5,5F10.3)
C.... RANDOMLY CHOOSE TRAINING SET
      DO 20 I=1,NTOT
   20 ITEMP(I) = I
      DO 30 I=1,NTRSET
   25 ITEST = 100.0*RAND(IX)
      IF (ITEMP(ITEST).EQ.0) GO TO 25
      IDTR(I) = ITEST
      ITEMP(ITEST) = 0
   30 CONTINUE
C.... REMAINING PATTERNS FORM PREDICTION SET
      II = 0
      DO 40 I=1,NTOT
      IF (ITEMP(I).EQ.0) GO TO 40
      II = II + 1
      IDPR(II) = I
   40 CONTINUE
      WRITE (NOUT,41) (IDPR(I),I=1,NPRSET)
   41 FORMAT (/,' PREDICTION SET MEMBERS',/,(' ',5I5))
C.... INITIALIZE WEIGHT VECTOR
      DO 50 J=1,NUM
   50 W(J) = 0.0
      W(NUM+1) = 1.0
      CALL TRAIN (DATA,W,LIST,NTRSET,NUM,NPASS,TSHD,NCONV,IDTR)
C.... CALL PREDICTION ROUTINE WITH THRESHOLD OF 0.75
      CALL PRED (DATA,LIST,W,NUM,TSHD,NPRSET,IDPR)
C.... CALL PREDICTION ROUTINE WITH THRESHOLD OF 0.0
      TSHD = 0.0
      CALL PRED (DATA,LIST,W,NUM,TSHD,NPRSET,IDPR)
      STOP
      END
```

```
      SUBROUTINE TRAIN (DATA,W,LIST,NTRSET,NUM,NPASS,TSHD,NCONV,IDTR)
C....
C.... IMPLEMENTS ERROR-CORRECTION FEEDBACK LINEAR LEARNING MACHINE
C....
      DIMENSION DATA(5,100),W(6),NSS(100),KPNT(20),LIST(100),IDTR(100)
      COMMON /IOUNIT/ NINP,NOUT
      NCONV = 0
      WRITE (NOUT,2)
    2 FORMAT (/,10X,'TRAINING ROUTINE')
      NUMM = NUM + 1
      NF = 0
      KNK = 0
      KNV = 0
C.... START OF MAIN LOOP (RETURN FROM STMT 206)
   51 KKK = 0
      IF (KNV) 54,54,53
   53 NDSS = KNV
      GO TO 65
   54 NDSS = NTRSET
      DO 60 I=1,NTRSET
   60 NSS(I) = IDTR(I)
C.... THE 200 LOOP CLASSIFIES THE NDSS MEMBERS OF THE CURRENT SUBSET
   65 DO 200 IR=1,NDSS
      I = NSS(IR)
C.... THE 70 LOOP CALCULATES THE DOT PRODUCT BETWEEN THE PATTERN
C....      AND THE CURRENT WEIGHT VECTOR
      S = W(NUMM)
      DO 70 J=1,NUM
   70 S = S + DATA(J,I)*W(J)
C.... NEXT THREE STMTS TEST FOR THE CORRECT ANSWER
      IF (LIST(I)) 95,95,96
   95 IF (S+TSHD) 200,200,116
   96 IF (S-TSHD) 115,115,200
C.... CALCULATE THE CORRECTION INCREMENT C
  115 C = 2.0*(TSHD-S)
      GO TO 117
  116 C = 2.0*(-TSHD-S)
  117 XX = 1.0
      DO 120 J=1,NUM
  120 XX = XX + DATA(J,I)**2
      C = C/XX
C.... THE 130 LOOP CORRECTS THE WEIGHT VECTOR
      DO 130 J=1,NUM
  130 W(J) = W(J) + C*DATA(J,I)
      W(NUMM) = W(NUMM) + C
      KKK = KKK + 1
      NSS(KKK) = I
      NF = NF + 1
  200 CONTINUE
      KNV = KKK
      KNK = KNK + 1
      KPNT(KNK) = KNV
      IF (KNK - 20) 205,203,203
  203 WRITE (NOUT,204) KPNT
  204 FORMAT (' ',20I3)
      KNK = 0
C.... STMT 205 TESTS FOR EXCESS NUMBER OF FEEDBACKS
  205 IF (NF-NPASS) 206,211,211
C.... STMT 206 TESTS WHETHER CURRENT SUBSET IS ENTIRE TRAINING SET
  206 IF (NDSS-NTRSET) 51,207,51
C.... STMT 207 TESTS WHETHER ZERO ERRORS WERE COMMITTED
  207 IF (KNV) 51,212,51
  211 NCONV = 1
C.... SUMMARY OUTPUT OF TRAINING ROUTINE
```

```
212 IF (KNK.GT.0) WRITE (NOUT,204) (KPNT(K),K=1,KNK)
    WRITE (NOUT,213) (W(J),J=1,NUMM)
213 FORMAT (/,10X,'WEIGHT VECTOR',/,(' ',F17.3))
    WRITE (NOUT,214) NF
214 FORMAT (/,10X,'FEEDBACKS',I6)
    RETURN
    END



    SUBROUTINE PRED (DATA,LIST,W,NUM,TSHD,NPRSET,IDPR)
C....
C.... PREDICTION ROUTINE
C....
    DIMENSION DATA(5,100),W(6),LIST(100),IDPR(100)
    COMMON /IOUNIT/ NINP,NOUT
    LW1 = 0
    LW2 = 0
    KW = 0
    NPA = 0
    NNA = 0
    DO 120 II=1,NPRSET
    I = IDPR(II)
    S = W(NUM+1)
    DO 50 J=1,NUM
 50 S = S + DATA(J,I)*W(J)
    IF (ABS(S)-TSHD) 101,102,102
101 KW = KW + 1
    GO TO 120
102 IF (LIST(I)) 103,103,105
103 NNA = NNA + 1
    IF (-S-TSHD) 104,104,120
104 LW1 = LW1 + 1
    GO TO 120
105 NPA = NPA + 1
    IF (S-THSD) 106,106,120
106 LW2 = LW2 + 1
120 CONTINUE
    WRITE (NOUT,121) TSHD
121 FORMAT (//,' PREDICTION WITH THRESHOLD =',F7.2)
    LWT = LW1 + LW2
    JW = NPA + NNA
    PW = 100.0-FLOAT(LWT)/FLOAT(JW)*100.0
    PW1 = 100.0-FLOAT(LW1)/FLOAT(NNA)*100.0
    PW2 = 100.0-FLOAT(LW2)/FLOAT(NPA)*100.0
    WRITE (NOUT,122) JW,KW,LWT
122 FORMAT (/,I10,' NUMBER PREDICTED',/,I10,
   X ' NUMBER NOT PREDICTED',/,I10,
   X ' NUMBER PREDICTED INCORRECTLY')
    WRITE (NOUT,123)
123 FORMAT (/,12X,'OVERALL',13X,'NO CLASS',13X,'YES CLASS')
    WRITE (NOUT,124) LWT,JW,PW,LW1,NNA,PW1,LW2,NPA,PW2
124 FORMAT (3(I10,'/',I3,1X,F6.2))
    RETURN
    END
```

```
      FUNCTION RAND (IX)
C.... PORTABLE FORTRAN RANDOM NUMBER GENERATOR
C....    FROM A.C.M. TRANS. MATH. SOFTWARE, 5, #2, 132 (1979)
C....    BY LINUS SCHRAGE
C....    USES THE RECURSION   IX = IX * A (MOD P)
C....
C.... INITIALIZE WITH SEED 0 < SEED < 2**31-1
C....
C.... USE EITHER RAND:  0 < RAND < 1
C....           OR IX:  0 < IX < 2**31 -1
C....
C.... CHECKING VALUES: IF IX(0)=1, THEN IX(1000)=522329230
C....
C.... IX IN CALLING LIST MUST BE INTEGER*4 IN CALLING PROGRAM
C....
      INTEGER A,P,IX,B15,B16,XHI,XALO,LEFTLO,FHI,K
C.... 7**5,2**15,2**16,2**31-1
      DATA A/16807/,B15/32768/,B16/65536/,P/2147483647/
C....
C.... GET 15 HIGH ORDER BITS OF IX
      XHI = IX/B16
C.... GET 16 LOW BITS OF IX AND FORM LO PRODUCT
      XALO = (IX-XHI*B16)*A
C.... GET 15 HIGH ORDER BITS OF LO PRODUCT
      LEFTLO = XALO/B16
C.... FORM THE 31 HIGHEST BITS OF FULL PRODUCT
      FHI = XHI*A + LEFTLO
C.... GET OVERFLOW PAST 31ST BIT OF FULL PRODUCT
      K = FHI/B15
C.... ASSEMBLE ALL THE PARTS AND PRESUBTRACT P
      IX = (((XALO-LEFTLO*B16) - P) + (FHI-K*B15)*B16) + K
C.... ADD P BACK IN IF NECESSARY
      IF (IX.LT.0) IX = IX + P
C.... MULTIPLY BY 1/(2**31-1)
      RAND = FLOAT (IX) * 4.656612875E-10
      RETURN
      END
```

LEARNING MACHINE PROGRAM

PREDICTION SET MEMBERS

2	3	11	14	15
16	19	20	28	38
39	45	46	48	58
62	65	77	79	100

TRAINING ROUTINE

```
29 14  8  5  3  2  2  2  2  2  1  0 10  4  4  2  1  0  6
 3  3  1  0  8  5  1  0  9  5  5  5  4  4  4  4  4  4  4
 4  4  3  3  3  3  3  3  2  2  2  2  0  8  3  3  2  1  0  7
 4  4  3  2  2  0  6  5  4  4  4  4  4  3  2  1  0  4  2
 2  2  0  6  3  3  0  3  0  7  7  7  6  6  6  6  6  6  6
 6  6  5  5  5  5  5  5  5  5  5  5  5  5  5  5  5  5  5
 4  4  4  4  4  4  4  3  3  3  3  3  2  1  0  0
```

WEIGHT VECTOR
```
     0.573
     1.187
     0.605
    -0.971
    -0.544
     0.902
```

```
     FEEDBACKS    532

PREDICTION WITH THRESHOLD =    0.75

      17  NUMBER PREDICTED
       3  NUMBER NOT PREDICTED
       0  NUMBER PREDICTED INCORRECTLY

          OVERALL            NO CLASS            YES CLASS
      0/ 17 100.00       0/  3 100.00       0/ 14 100.00


PREDICTION WITH THRESHOLD =    0.00

      20  NUMBER PREDICTED
       0  NUMBER NOT PREDICTED
       0  NUMBER PREDICTED INCORRECTLY

          OVERALL '           NO CLASS            YES CLASS
      0/ 20 100.00       0/  6 100.00       0/ 14 100.00
```

REFERENCES

Aishima, T., and S. Nakai, "Chemometrics in Flavor Research," *Food Rev. Internatl.* **7**, 33 (1991).

Brown, S. D., "Chemometrics," *Anal. Chem.* **62**, 84R (1990).

Brown, S. D., T. Q. Barker, R. J. Larivee, S. L. Monfre, and H. R. Wilk, "Chemometrics," *Anal. Chem.* **60,** 242R (1988).

Brown, S. D., R. S. Bear, Jr., and T. B. Blank, "Chemometrics," *Anal. Chem.* **64**, 22R (1992).

Brunner, T. R., C. L. Wilkins, R. C. Williams, and P. J. McCombie, "Pattern Recognition Analysis of Carbon-13 Free Induction Decay Data," *Anal. Chem.,* **47**, 662–665 (1975).

Burgard, D. R., S. P. Perone, and J. L. Wiebers, "Sequence Analysis of Oligodeoxyribonucleotides by Mass Spectrometry. 2. Application of Computerized Pattern Recognition to Sequence Determination of Di-, Tri-, and Tetranucleotides," *Biochemistry,* **16**, 1051–1057 (1977).

Clark, H. A., and P. C. Jurs, "Classification of Crude Oil Gas Chromatograms by Pattern Recognition Techniques," *Anal. Chem.,* **51**, 616–623 (1979).

de Jong, S., "Chemometrical Applications in an Industrial Food Research Laboratory," *Mikrochim. Acta* **II**, 93 (1991).

Delaney, M. F., "Chemometrics," *Anal. Chem.,* **56**, 261R–277R (1984).

Duewer, D. L., and B. R. Kowalski, "Forensic Data Analysis by Pattern Recognition. Categorization of White Bond Papers by Elemental Composition," *Anal. Chem.,* **47**, 526–530 (1975).

Edlund, U., and S. Wold, "Interpretation of NMR Substituent Parameters by the Use of a Pattern Recognition Approach," *J. Magn. Res.,* **37**, 183 (1980).

Forina, M., and E. Tiscornia, "Pattern Recognition Methods in the Prediction of Italian Olive Oil Origin hy Their Fatty Acid Content," *Ann. Chim.,* **72**, 143–155 (1982).

Frank, I. E., and B. R. Kowalski, "Chemometrics," *Anal. Chem.*, **54**, 232R–243R (1982).

Gaarenstroom, P. D., S. P. Perone, and J. L. Moyers, "Application of Pattern Recognition and Factor Analysis for Characterization of Atmospheric Particulate Composition in Southwest Desert Atmosphere," *Environ. Sci. Tech.*, **11**, 795–800 (1977).

Gemperline, P. J., K. H. Miller, T. L. West, J. E. Weinstein, J. C. Hamilton, and J. T. Bray, "Principal Component Analysis, Trace Elements, and Blue Crab Shell Disease," *Anal. Chem.* **64**, 523A (1992).

Hancock, D. O., and R. E. Synovec, "Early Detection of C-130 Aircraft Engine Malfunction by Principal Component Analysis of the Wear Metals in C-130 Engine Oil," *Appl. Spectr.* **43**, 202 (1989).

Hartigan, J. A., *Clustering Algorithms,* Wiley, New York, 1975.

Hayek, E. W. H., P. Krenmayr, H. Lohninger, U. Jordis, W. Moche, and F. Sauter, "Identification of Archaeological and Recent Wood Tar Pitches Using Gas Chromatography/Mass Spectrometry and Pattern Recognition," *Anal. Chem.* **62**, 2038 (1990).

Henry, D. R., P. C. Jurs, and W. A. Denny, "Structure-Antitumor Activity Relationships of 9-Anilinoacridines Using Pattern Recognition," *J. Med. Chem.*, **25**, 899–908 (1982).

Huber, J. F. K., and G. Reich, "Characterization and Selection of Stationary Phases for Gas-Liquid Chromatography by Pattern Recognition Methods," *J. Chromatogr.*, **294**, 15 (1984).

Jain, A. K., and R. C. Dubes, *Algorithms for Clustering Data,* Prentice-Hall, Englewood Cliffs, NJ, 1988.

Jellum, E., I. Bjoernson, R. Nesbakken, E. Johansson, and S. Wold, "Classification of Human Cancer Cells by Means of Capillary Gas Chromatography and Pattern Recognition Analysis," *J. Chromatogr.*, **217**, 231–237 (1981).

Jurs, P. C., and T. L. Isenhour, *Chemical Applications of Pattern Recognition,* Wiley-Interscience, New York. 1975.

Jurs, P. C., C. L. Ham, and W. E. Brugger, "Computer Assisted Studies of Chemical Structure and Olfactory Quality Using Pattern Recognition Techniques," *A.C.S. Symp. Ser.*, **148**, 143 (1981).

Jurs, P. C., B. R. Kowalski, T. L. Isenhour, and C. N. Reilley, "Computerized Learning Machines Applied to Chemical Problems. Molecular Structure Parameters from Low Resolution Mass Spectrometry," *Anal. Chem.*, **42**, 1387–1394 (1970).

Kirschner, G. L., and B. R. Kowalski, "The Application of Pattern Recognition to Drug Design," in *Drug Design,* Vol. 8, E. J. Ariens (ed.), Academic Press, New York, 1979.

Kowalski, B. R., "Chemometrics," *Anal. Chem.*, **52**, 112R–122R (1980).

Kowalski, B. R., and C. F. Bender, "The Application of Pattern Recognition to Screening Prospective Anticancer Drugs. Adenocarcinoma 755 Biological Activity Test," *J. Am. Chem. Soc.*, **96**, 916–918 (1974).

Kowalski, B. R., and S. Wold, "Pattern Recognition in Chemistry," in *Handbook of Statistics,* Vol. 2, P. R. Krishnaiah and L. N. Kanal (eds.), North-Holland, New York, 1982.

Kryger, L., "Interpretation of Analytical Chemistry Information by Pattern Recognition Methods: A Survey," *Talanta,* **28**, 871–887 (1981).

Kwan, W. O., and B. R. Kowalski, "Correlation of Objective Chemical Measurements and Subjective Sensory Evaluations. Wines of *Vitis Vinifera* Variety 'Pinot Noir' from France and the United States," *Anal. Chim. Acta,* **122**, 215–222 (1980).

Lavine, B. K., and D. A. Carlson, "Chemical Fingerprinting of Africanized Honeybees by Gas Chromatography/Pattern Recognition Techniques," *Microchem. J.* **42**, 121 (1990).

Massart, D. L., B. G. M. Vandeginste, S. N. Deming, Y. Michotte, and L. Kaufman, *Chemometrics: A Textbook,* Elsevier, Amsterdam, 1987.

Massart, D. L., L. Kaufman, and K. H. Esbensen, "Hierarchical Nonhierarchical Clustering Strategy and Application to Classification of Iron Meteorites According to Their Trace Element Patterns," *Anal. Chem.,* **54**, 911 (1982).

McGill, J. R., and B. R. Kowalski, "Recognizing Patterns in Trace Elements," *Appl. Spectrosc.,* **31**, 87–95 (1977).

Menon, G. K., and A. Cammarata, "Pattern Recognition II: Investigation of Structure-Activity Relationships," *J. Pharm. Sci.,* **66**, 304–314 (1977).

Miyashita, Y., A. Li, and S. Sasaki, "Chemical Pattern Recognition and Multivariate Analysis for QSAR Studies," *Trends Anal. Chem.* **12**, 50 (1993).

Nilsson, N. J., *Learning Machines,* McGraw-Hill, New York, 1965.

Norden, B., U. Edlund, and S. Wold, "Carcinogenicity of Polycyclic Aromatic Hydrocarbons Studied by SIMCA Pattern Recognition," *Acta Chem. Scand.,* **B32**, 602–608 (1978).

Ramos, L. S., K. R. Beebe, W. P. Carey, E. Sanchez, B. C. Erickson, B. E. Wilson, L. E. Wangen, and B. R. Kowalski, "Chemometrics," *Anal. Chem.* **58**, 294R (1986).

Raznikov, V. V., and V. L. Talroze, *Dokl. Akad. Nauk SSSR,* **170**, 379 (1966).

Rhodes, G., M. Miller, M. L. McConnell, and M. Novotny, "Metabolic Abnormalities Associated with Diabetes Mellitus, as Investigated by Gas Chromatography and Pattern Recognition Analysis of Profiles of Volatile Metabolites," *Clin. Chem.,* **27**, 580–585 (1981).

Ritter, G. L., S. R. Lowry, C. L. Wilkins, and T. L. Isenhour, "Simplex Pattern Recognition," *Anal. Chem.,* **47**, 1951 (1975).

Rose, S. L., and P. C. Jurs, "Computer-Assisted Studies of Structure Activity Relationships of N-Nitroso Compounds Using Pattern Recognition," *J. Med. Chem.,* **25**, 769–776 (1982).

Rotter, H., and K. Varmuza, "Computer-Aided Interpretation of Steroid Mass Specira by Pattern Recognition Methods. Part III. Computation of Binary Classifiers by Linear Regression," *Anal. Chim. Acta,* **103**, 61–71 (1978).

Senn, P., "Cluster Analysis for Chemists," *Tetra. Comp. Meth.* **2**, 133 (1989).

Sharaf, M. A., D. A. Illman, and B. R. Kowalski, *Chemometrics,* Wiley, New York, 1986.

Shattuck, T. W., M. S. Germani, and P. R. Buscek, "Cluster Analysis of Chemical Compositions of Individual Atmospheric Particles Data," in *Environmental Applications of Chemometrics,* J. J. Breen (ed.), American Chemical Society, Washington, DC, 1985.

Stouch, T. R., and P. C. Jurs, "Chance Factors in Nonparametric Linear Discriminant Studies," *Quant. Struct.-Act. Relat.* **5**, 57 (1986).

Stuper, A. J., W. E. Brugger, and P. C. Jurs, *Computer Assisted Studies of Chemical Structure and Biological Function,* Wiley-Interscience, New York, 1979.

Tou, J. T., and R. C. Gonzalez, *Pattern Recognition Principles,* Addison-Wesley, Reading, MA, 1974.

Varmuza, K., *Pattern Recognition in Chemistry,* Springer-Verlag, Berlin, 1980.

Varmuza, K., "Some Aspects of the Application of Pattern Recognition Methods in Chemistry," in *Computer Applications in Chemistry,* S. R. Heller and R. Potenzone (eds.), Elsevier Scientific, Amsterdam 1983.

Wang, P. H. (ed.), *Graphical Representation of Multivariate Data,* Academic, New York, 1978.

Wilkins, C. L., and P. C. Jurs, "Fourier and Hadamard Transforms in Pattern Recognition," in *Transform Techniques in Chemistry,* P. R. Griffiihs (ed.), Plenum, New York, 1978, pp. 307–331.

Willett, P., *Similarity and Clustering in Chemical Information Systems,* Research Studies Press, Letchworth, England, 1987.

Willett, P., V. Winterman, and D. Bawden, "Implementation of Nonhierarchic Cluster Analysis Methods in Chemical Information Systems: Selection of Compounds for Biological Testing and Clustering of Substructure Search Output," *J. Chem. Inf. Comp. Sci.,* **26**, 109 (1986).

Wold, S., "Pattern Recognition by Means of Disjoint Principal Components Models," *Pattern Recogn.,* **8**, 127 (1976).

Wold, S., and M. Sjostrom, "A Method for Analyzing Chemical Data in Terms of Similarity and Analogy," *A.C.S. Symp. Ser.,* **52**, 243–282 (1977).

Woodruff, H. B., and M. E. Munk, "Computer-Assisted Interpretation of Infrared Spectra," *Anal. Chim. Acta,* **95**, 13–23 (1977).

Zupan, J., "Hierarchical Clustering of Infrared Spectra," *Anal. Chim. Acta,* **139**, 143 (1982).

15

ARTIFICIAL INTELLIGENCE AND EXPERT SYSTEMS

Artificial intelligence (AI) is a field of research in which the following goals are pursued:

1. To understand the nature of intelligence using the paradigm of information processing
2. To construct theories and build systems through programming computers that exhibit intelligent behavior requiring reasoning and perception
3. To render computers more useful by making them exhibit intelligent behavior

The central activity involved in intelligent behavior is information processing. Intelligence in human behavior is associated with tasks such as understanding natural language, learning, reasoning, solving problems, and playing complex games such as chess or "go." All of these areas have been studied with AI. Problem-solving programs have been developed in chemistry, biology, geology, engineering, and medicine that can perform tasks at a level similar to that of human experts. AI is practiced as an experimental science; that is, the construction of computer programs and demonstration of their capabilities are considered to be a vital part of the research.

Artificial intelligence has been a field of research for only about 30 years. Its underpinnings were advances in mathematical logic during the 1930s and 1940s and advances in computational capability during the 1940s and early 1950s. The development of electronic digital computers made possible the implementation of tasks that involved intelligence for the first time. An early landmark paper was by Turing (1950), who addressed many of the fundamental questions of intelligent behavior of machines and proposed what has come to be called the *Turing test*. Researchers were able to build systems to test their theories of intelligent behavior, leading to the typical scientific cycle of theory, experiment, theory revision, and so on. Progress was rapid. The early research focused on game-playing programs as a domain for study. Soon other domains were studied, including theorem proving, language understanding, vision, and speech recognition. Since the 1980s, AI has been a rapidly growing field with all the hallmarks of a relatively mature scientific discipline: a professional society of its own (American Association for Artificial Intelligence), both scientific and popular journals, textbooks, university courses, proponents, opponents, and so on (Duda and Shortliffe, 1983).

There is a deeply rooted *paradox* in AI. On one hand, computers are completely inflexible, they do slavishly what they are told to do, and they never deviate from following instructions exactly. On the other hand, the hallmark of intelligence is flexibility, adaptability, and the ability to respond or decide as the situation suggests. The conciliation of computer characteristics with the demands of intelligent behavior is an accomplishment and a goal of AI research. This aspect of AI has been discussed at length by Simon (1984) and Hofstadter (1979).

Intelligent behavior is developed in computers through implementation of software, that is, the programs. The programs execute on computer hardware, but that fact is relatively unimportant. It is the symbol manipulation done by the programs that is relevant, not the details of how the electronic circuits accomplish it.

There are two basic ingredients in intelligent behavior: search and knowledge. *Search* refers to the ability to create a space of possibilities that is large enough to contain the sought solution to the problem and then searching for that solution. Often the spaces for real problems increase in size very quickly as a function of the size of the problem. This is true, for example, in the looking ahead in a chess game for possible moves. There are something like 10^{120} sequences of legal moves in a chess game. It is beyond the power of any computer to examine them all and thereby exhaustively generate the perfect response move in chess. Such combinatorial explosions of space size often form a fundamental limitation on the capabilities of intelligent programs.

The second basic ingredient of intelligent behavior is the possession of *knowledge*. A popular phrase of recent AI research is "Knowledge is power." AI applications make great demands on the knowledge bases employed. The knowledge is usually diverse and interrelated. The knowledge can be used to guide the search in order to limit the number of possible solutions that must be examined. The representation of the knowledge must be effective, so that the knowledge can be used by the system to attain its goals. The organization of the knowledge so that it is accessible and can be found when needed is a very important issue in AI research. Several different paradigms have been developed for the organization of the knowledge bases in AI programs. One is standard symbolic logic, such as "All fish swim." Another representation is that of production systems that use if–then rules, which will be described below. Specialized representations such as frames or scripts have been developed as well. In addition to the organization of the knowledge, practical AI systems must provide a means for acquiring the knowledge and inserting it into the knowledge base. The automation of knowledge acquisition is a research area within AI. The next step of knowledge acquisition, learning from experience or examples, is a newly emerging area of AI research.

An example based on an idea of Rich (1984) will illustrate the different demands made of knowledge bases by AI programs as opposed to ordinary programs. Take as an example of a normal database the information contained in an ordinary almanac. Facts abound in such a database, often presented in tabular form. This type of information can be stored and manipulated by computer very easily. Questions based on such facts can be answered by straightforward methods. The answers to questions of the following types can be looked up in the appropriate part of the database: "What is the tallest mountain in the world?" or "Who was the first secretary general of the United Nations?" Now, suppose that the following fact is contained in the almanac: Sacramento is the capital of California. On the basis of this isolated fact, a person would be able to answer any one of the following questions: "Is San Diego the capital of California?" What state is Sacramento in?" "Where is the government of California?" "Is Sacramento in the United States?" The answers to these questions are not contained in the simple piece of factual information given. Additional information is needed to answer these questions, information such as: states have just one capital; a capital of a state is in that state; if a is in b, and b is in c, then a is in c; California is a state; a state has its government in its capital; the United States is composed of states; and so on. The kind of knowledge manipulation required to put facts together in a flexible way is one key to AI program success.

One important class of AI programs consists of *expert systems* that are designed to serve as consultants for decision making. The fundamental aspect of intelligence that AI systems must implement is how to represent large amounts of information in a way that allows for its effective use. For a program to exhibit intelligence, it must have access to large amounts of knowledge and must know how to manipulate it and use it. The manipulations are sometimes general, but they are often specific to the domain of interest. This approach to the development of expert systems has been called the *knowledge-based approach.*

Expert systems have been developed within a number of narrow domains. The knowledge bases have been acquired from human experts, who are experts precisely because of their extensive knowledge within their domain. An important characteristic of expert systems is that they are not limited to the knowledge of one person but can store and use the rules as known to a group of community of experts. One characteristic of this knowledge that must be captured in an expert system is that the knowledge is not always certain or objective. It can be subjective, judgmental, or even rules of thumb. The type of information used and the type of processing that is done with it often does not conform to the definition of *algorithm*. Rather, the correct labeling is *heuristic.* A heuristic is a rule or way of proceeding that is usually all right but may not always be optimum. Several examples of heuristics are as follows: control of the center of the board is important in chess; a falling barometer means that it very well may rain soon; red sky at night is sailor's delight; a florid complexion may mean high blood pressure. Expert systems must deal with their knowledge base of information through the use of heuristics in attempting to make decisions.

The most common form for representation of information is that of production rules, or if-then rules. They have the following form

IF (premise) THEN (action)

where the conditions with (premise) can be very complicated and the (action) can also contain a series of subactions. Ordinarily, only human experts have the knowledge of the proper pairs of premises and actions.

The basic research issues within AI are knowledge acquisition, knowledge representation, inference and uncertainty, and explanation. The current bottleneck in the development of expert systems is the acquisition and representation of the knowledge base.

Knowledge Acquisition. Domain knowledge must be extracted from the human experts. In the development of current expert systems, this has largely been done manually by collaboration between an AI researcher and a domain expert. This has meant having computer scientists interview experts in the domain of applications. The knowledge acquired in this way usually is in the form of English sentences. It must be structured by the computer scientist so that it can be properly represented by the computer system. This task has come to be called *knowledge engineering.* The slow going of this manual method has led many to seek methods that would automate knowledge acquisition or, better yet, allow systems to learn from examples presented in a natural way. This will remain a goal for some time, however.

Knowledge Representation. Knowledge must be represented inside the expert system so that it can be used effectively. The representation must be accessible and flexible. The knowledge must be represented so that it can be modified and augmented, and the knowledge must be represented so that the expert system can explain its actions to humans.

Inference and Uncertainty. Decisions are made in expert systems by weighing evidence, taking into account the (possibly many) rules that have bearing on the question at hand, and sometimes calculating probabilities. However, the most effective methods for incorporating uncertainty in the validity of the rules or the probabilities of correct application as a function of the situation have not been fully defined as yet.

Explanation. This refers to the necessity for expert systems to be able to provide their reasoning sequence to human users so that they can under-

correlations, largely through the definition of 18 specific classes of mass spectral data. These classes were chosen because they are indicative of a variety of structural entities.

When STIRS is interpreting an unknown spectrum, it matches the data in each class for the unknown to the data in each class for each reference spectrum. It computes a match factor that is a quantitative measure of the degree of similarity of the unknown spectrum and each reference spectrum. For each class, the set of 15 best-matched reference spectra are saved for further analysis. The molecular structures of the compounds corresponding to each of the best-matched spectra are compared to seek common substructures. If a substructure is found to be present in many of the retrieved structures, it is probable that this substructure will also be present in the unknown compound.

STIRS can also predict the molecular weight of the unknown compound. The approach is based on the assumption that the best-matching compounds selected from the reference file will have primary neutral losses that are similar to those of the unknown compound. The program uses the mass values at the upper end of the unknown mass spectrum to compare to the upper mass values in the best-matched compounds. The predictions were reported to be better than 90% correct in one test of the system (Mun et al. 1981).

16.2   INFRARED SPECTRA

The infrared spectrum of a compound provides structural information that is complementary to that contained in the mass spectrum. The individual IR absorptions in the spectrum are due to the bonding configurations present in the structure. A carbonyl group gives rise to a signature absorption, as one specific example. IR provides a method for "fingerprinting" organic compounds with regard to their functional groups. The presence of certain atom types, certain bond types, and functional groups can be confirmed by comparing the observed spectrum with correlation charts. The rules contained in the correlation charts have been built up over many years by experimentalists through the codification of empirical observations. Infrared spectra are commonly used in conjunction with other forms of spectroscopy for the elucidation of molecular structures of unknown compounds.

The most desirable way to use an infrared spectrum of an unknown compound is to compare it to the members of a reference file of identified spectra. A strong resemblance between the IR spectrum of the unknown and one of the file spectra is strong evidence of structural similarity or even identity. This searching option is only available if one has access to a large file or identified spectra of similar structural types to the unknown. Such collections of IR spectra are costly to generate and maintain, so this means of interpreting the IR spectrum of an unknown is seldom feasible. There are, as noted above, several collections of IR spectra now in existence, and library searching software allows their use for this task.

PAIRS

The alternative to library searching is interpretation of the IR spectrum by comparing it to empirical correlation charts that relate infrared absorption bands to structural units. Quite extensive sets of empirical data are available in reference volumes. Automation of the reasoning process used by chemists to interpret infrared spectra has been attempted by several workers. Work started at in the late 1970s (Woodruff and Munk 1977) and continued through the 1980s (Woodruff and Smith 1980, 1981; Tomellini et al. 1981, 1984a, 1984b; Woodruff 1984) has yielded a computer program called PAIRS (program for the analysis of infrared spectra). The original version ran on an IBM mainframe computer, and a version was developed to run on a Nicolet 1180 laboratory minicomputer. A more recent version, with substantial improvements, has been made available for VAX computers as QCPE Program No. 497 (Tomellini et al. 1985).

PAIRS does not use or store a database of IR spectra, and it is therefore not subject to the limitations associated with all library search procedures. Woodruff (1984) stated: "PAIRS attempts to parallel as closely as possible the reasoning a spectroscopist uses in interpreting IR spectra."

PAIRS consists of several parts that interact with each other. The two most important parts of the system are (1) the set of empirical infrared spectroscopy rules that form a base of factual information for the program to use and (2) the methodology for manipulating the rules to obtain overall probabilities to report to the user. The interpretation rules are treated by the program as data, which is to say that they can be changed, updated, and corrected without the need to alter the program itself in any way. This approach to organizing a program is used routinely in expert systems, and it has been purposely employed in the development of the PAIRS system because of its great enhancement to program understanding and updating. The rules are input to the data base by the chemist using a specially designed language called CONCISE (computer-oriented notation concerning IR spectral evaluation). The language was designed to make the input of the IR spectroscopy rules as natural as possible to the spectroscopist.

When used for interpretation, PAIRS first requires the user to input the IR spectrum to be interpreted in digital format. The spectrum is digitized over the range 4000–5000 cm^{-1} with peak intensities coded from 1 (for very weak) to 10 (for very strong) and with peak widths of 1 (sharp), 2 (average), or 3 (broad). Ancillary information such as the sample state and the molecular formula are then input. The program then uses the rule set to perform the interpretation. Its output consists of probabilities for each functional group that is under consideration.

The design and construction of the rule set is the heart of PAIRS, and it is this aspect of the overall design of PAIRS that we will discuss in the following paragraphs. The rules used by PAIRS are set up the same way that spectroscopists would set them up, that is, by spectral regions that correspond to specific functional groups. Each functionality defines a class. The rules for each functionality are set up in the form of a hierarchical decision tree. Definite regions of the spectrum are examined sequentially. When an absorption is found in a region being examined, the conclusion reached is dictated by the way in which the tree is set up. The decision trees are related to the functionality. A specific example taken from Woodruff and Smith (1981) will clarify this method for representing IR spectral information.

The example used by Woodruff and Smith was aldehydes. The information regarding absorptions that correlate with the presence of the aldehyde functionality is first found in standard reference books. The consensus of the rules for aldehydes is as follows: (1) a strong peak due to carbonyl appears between 1765 and 1660 cm^{-1}, and its exact position depends on the immediate surroundings of the carbonyl group; and (2) two peaks of moderate intensity appear between 2900 and 2695 cm^{-1}, usually near 2820 and 2720 cm^{-1}. The 2720-cm^{-1} peak is usually sharp and resolved from interferences; it provides better evidence for aldehyde than the 2820-cm^{-1} peak.

These rules for aldehydes can be expressed as a decision tree, as shown in Figure 16.1. The entry point for this tree is at the left. The initial probability of an aldehyde moiety being present is zero. At each junction in the tree, a question is answered either "yes" or "no," and the corresponding branch of the tree is followed. The first question to be answered is "Does the IR spec-

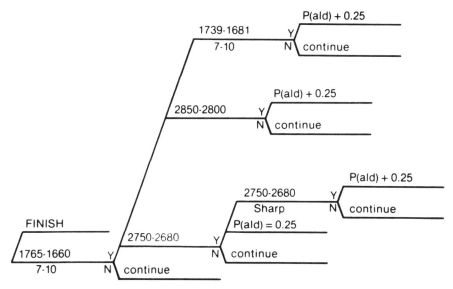

Figure 16.1 Decision tree expressing the interpretation rules for aldehyde compounds as used by PAIRS. (Reprinted with permission from *Progress in Industrial Microbiology,* Vol. 17, Elsevier Science Publishers, New York, 1983. Copyright 1983, Elsevier Science Publishers.)

trum contain a strong absorption in the range 1765–1660 cm^{-1}?" If not, then the examination of the tree is finished, and PAIRS would continue to another section of its decision tree network (this is what "continue" means). If the peak is found, the Y branch is taken, and the region from 2750 to 2680 cm^{-1} is examined. If no peak is found, the aldehyde decision tree is abandoned. If a peak is found, the probability of aldehyde is set to 0.25. If the peak in this region is sharp, the probability of aldehyde is increased by 0.25. Then the region from 2850 to 2800 cm^{-1} is examined; if a peak is found in this region, the probability of aldehyde is increased by 0.25. Finally, the region from 1739 to 1681 cm^{-1} is examined, and if a strong peak is found, the probability of aldehyde is increased by 0.25. If all four tests were positive, the probability of aldehyde would be 1.0. This is to say that if all four aldehyde tests are passed, then the system is certain that an aldehyde moiety is present.

When these simple rules for aldehyde functionality were applied to real IR spectra, the results were not very satisfactory. However, these rules are quite simple and are used as an illustration of how the spectral information is incorporated into PAIRS. Woodruff and Smith (1981) present a much more complicated and complete decision tree for aldehydes that indicates account whether the sample was run in mineral oil, whether an acid functionality might be present, and some other factors. When this enhanced decision tree for aldehydes was tested, the results of the interpretations were much improved, as would be expected.

The information present in the decision trees of PAIRS is entered by the chemist using CONCISE. Smith and Woodruff (1984) have published a detailed description of CONCISE. It is designed to allow the chemist to generate, update, and maintain a set of infrared rules without having to know any computer language. Moreover, PAIRS was designed so that whenever the system makes faulty interpretations, the user can inspect the rules used. Then they can be modified, if desirable. At the very minimum, the user will be able to see exactly why PAIRS made a faulty interpretation.

In interpreting an IR spectrum, PAIRS makes its way through all the decision trees available in its knowledge base. It then reports the probabilities of functional groups being present. In an early paper, Woodruff and Smith (1980) report that PAIRS had rules for more than 170 functional groups and functional group types. PAIRS considers organic compounds with atom types of C, H, O, N, S, F, Cl, and Br whose spectra have been taken in one of six sample states. A subsequent report on PAIRS described its modification to be able to deal with vapor-phase spectra (which require rules different from the original rules for condensed-phase spectra). PAIRS is able to deal with mixture samples just as easily as pure samples because no comparisons are done with reference spectra. Of course, if one of the components of a mixture is at a very low concentration, peaks would have lower than expected intensities, which could affect the results of the interpretation. The user must watch for effects of impurities on results.

PAIRS has been extensively tested with hundreds of IR spectra of complex molecules, and the results show that the concept being investigated is valid. That is, one can develop a computerized assistant to help the chemist in a complicated task such as interpreting IR spectra of organic compounds.

In more recent work, the PAIRS development group has reported the generation of programs that create IR interpretation rules in CONCISE based on representative spectra (Tomellini 1984a). The process is automated but also interactive. An automatic peak-finding routine provides information about peak positions, peak intensities, and peak widths. The program was developed using over 3300 vapor-phase IR spectra. The rule-generating capabilities of the program were tested by developing rules for saturated alcohols with a set of 51 vapor-phase spectra of 19 primary, 18 secondary, and 14 tertiary alcohols. Rules that could reliably differentiate between these three subclasses of alcohols were developed by the program. This work is an example of the automation of the knowledge acquisition part of an expert system.

PAIRS has been incorporated into a larger software system for identification of compounds present in environmental samples from hazardous-waste remedial-action sites (Ying et al. 1987). The identifications are made from the IR spectrum of the mixture. The method was evaluated using the 62 most commonly identified organic compounds found in hazardous-waste sites. The compounds present in test sets of four-component mixtures were identified with a very high degree of accuracy by the system.

REFERENCES

General

Abe, H., I. Fujiwara, T. Nishimura, T. Okuyama, T. Kidil, and S. Sasaki, "Recent Advances in the Structure Elucidation System, CHEMICS," *Comput. Enhanc. Spectrosc.,* **1**, 55–62 (1983).

Clerc, J. T., E. Pretsch, and J. Seibl, *Structural Analysis of Organic Compounds by Combined Application of Spectroscopic Methods,* Elsevier, Amsterdam, 1981.

Munk, M. E., C. A. Shelley, H. B. Woodruff, and M. O. Trulson, "Computer-Assisted Structure Elucidation," *Fres. Z. Anal. Chem.,* **313**, 473–479 (1982).

Sasaki, S., and Y. Kudo, "Structure Elucidation System Using Structural Information from Multisources: CHEMICS," *J. Chem. Inf. Comput. Sci.,* **25**, 252–257 (1985).

Shelley, C. A., and M. E. Munk, "CASE: A Computer Model of th Structure Elucidation Process," *Anal. Chim. Acta,* **133**, 507–516 (1981).

Smith, D. H., *Computer-Assisted Structure Elucidation, A.C.S. Symp. Ser.,* **54**, 1977.

Warr, W. A., "Spectral Databases," *Chemom. Intell. Lab. Syst.,* **10**, 279–292 (1991).

Warr, W. A., "Computer-Assisted Structure Elucidation. Part 1. Liberty Search and Spectral Data Collections," *Anal. Chem.,* **65**, 1045A (1993a).

Warr, W. A., "Computer-Assisted Structure Elucidation. Part 2. Indirect Database Approaches and Established Systems," *Anal. Chem.,* **65**, 1087A (1993b).

Mass Spectra and PBM and STIRS

Atwater (Fell), B. L., R. Venkataraghaven, and F. W. McLafferty, "Matching of Mixture Mass Spectra by Subtraction of Reference Spectra," *Anal. Chem.*, **51**, 1945–1949 (1979).

Dayringer, H. E., G. M. Pesyna, R. Venkataraghavan, and F. W. McLafferty, "Computer-Aided Interpretation of Mass Spectra. IX. Information on Substructural Probabilities from Self-Training Interpretive and Retrieval System (STIRS)," *Org. Mass Spectrom.*, **11**, 529–542 (1976).

Haraki, K. S., R. Venkataraghavan, and F. W. McLafferty, "Prediction of Substructures from Unknown Mass Spectra by the Self-Training Interpretive and Retrieval System," *Anal. Chem.*, **53**, 386–392 (1981).

Hertz, H. S., R. A. Bites, and K. Biemann, "Identification of Mass Spectra by Computer-Searching a File of Known Spectra," *Anal. Chem.*, **43**, 681–691 (1971).

Kwok, K-S., R. Venkataraghavan, and F. W. McLafferty, "Computer-Aided Interpretation of Mass Spectra. III. Self-Training Interpretative and Retrieval System," *J. Am. Chem. Soc.*, **95**, 4185–4194 (1973).

Lowry, S. R., T. L. Isenhour, J. B. Justice, Jr., F. W. McLafferty, H. E. Dayringer, and R. Venkataraghavan, "Comparison of Various K-Nearest Neighbor Voting Schemes with the Self-Training Interpretive and Retrieval System for Identifying Molecular Structures," *Anal. Chem.*, **49**, 1720–1722 (1977).

Martinsen, D. P., "Survey of Computer Aided Methods for Mass Spectral Interpretation," *Appl. Spectrosc.*, **35**, 255–266 (1981).

McLafferty, F. W., and D. B. Stauffer, "Interpretative Computer Programs for Mass Spectrometry," *J. Chem. Inf. Comput. Sci.*, **25**, 245–252 (1985).

McLafferty, F. W., R. H. Hertel, and R. D. Villwock, "Computer Identification of Mass Spectra. VI. Probability Based Matching of Mass Spectra. Rapid Identification of Specific Compounds in Mixtures," *Org. Mass Spectrom.*, **9**, 690–702 (1974).

McLafferty, F. W., S. Cheng, D. M. Dully, C.-J. Guo, I. K. Mun, D. W. Peterson, S. O. Russo, D. A. Salvucci, J. W. Serum, W. Staedeli, and D. B. Stauffer, "Matching Mass Spectra Against a Large Data Base During GC/MS Analysis," *Internatl. J. Mass Spectrom. Ion Phys.*, **47**, 317–319 (1977).

McLafferty, F. W., D. B. Stauffer, A. B. Twiss-Brooks, and S. Y. Loh, "An Enlarged Data Base of Electron-Ionization Mass Spectra," *J. Am. Soc. Mass Spectrom.*, **2**, 432–437 (1991).

Mun, I. K., D. R. Bartholomew, D. B. Stauffer, and F. W. McLafferty, "Weighted File Ordering for Fast Matching of Mass Spectra Against a Comprehensive Data Base," *Anal. Chem.*, **53**, 1938–1939 (1981).

Pesyna, G. M., R. Venkataraghavan, H. E. Dayringer, and F. W. McLafferty, "Probability Based Matching System Using a Large Collection of Reference Mass Spectra," *Anal. Chem.*, **48**, 1362–1368 (1976).

Venkataraghavan, R., H. E. Dayringer, G. M. Pesyna, B. L. Atwater, I. K. Mun, M. M. Cone, and F. W. McLafferty, "Computer-Assisted Structure Identification of Unknown Mass Spectra," in *Computer-Assisted Structure Elucidation*, D. H. Smith (ed.), American Chemical Society, Washington, DC, 1977.

Infrared Spectra and PAIRS

Smith, G. M., and H. B. Woodruff, "Development of a Computer Language and Compiler for Expressing the Rules of Infrared Spectral Interpretation," *J. Chem. Inf. Comput. Sci.,* **24**, 33–39 (1984).

Tomellini, S. A., D. D. Saperstein, J. M. Stevenson, G. M. Smith, H. H. Woodruff, and P. F. Seelig, "Automated Interpretation of Infrared Spectra with an Instrument Based Minicomputer," *Anal. Chem.,* **53**, 2367–2369 (1981).

Tomellini, S. A., J. M. Stevenson, and H. B. Woodruff, "Rules for Computerized Interpretation of Vapor-Phase Infrared Spectra," *Anal. Chem.,* **56**, 67–70 (1984a).

Tomellini, S. A., R. A. Hartwick, J. M. Stevenson, and H. B. Woodruff, "Automated Rule Generation for the Program for the Analysis of Infrared Spectra (PAIRS)," *Anal. Chim. Acta,* **162**, 227–240 (1984b).

Tomellini, S. A., R. A. Hartwick, and H. B. Woodruff, "Automatic Tracing and Presentations of Interpretation Rules Used by PAIRS," *Appl. Spectrosc.,* **39**, 331–333 (1985).

Tomellini, S. A., G. M. Smith, and H. B. Woodruff, "PAIRS: Program for the Analysis of Infrared Spectra (VAX Version)," QCPE Program No. 497.

Woodruff, H. B., "Progress in Interpretation of Antibiotic Structures Using Computerized Infrared Techniques," in *Progress in Industrial Microbiology,* Vol. 17, M. E. Hushell (ed.), Elsevier Scientific, Amsterdam, 1983, pp. 71–108.

Woodruff, H. B., "Using Computers to Interpret IR Spectra of Complex Molecules," *Trends Anal. Chem.* **3**, 72–75 (1984).

Woodruff, H. B., and M. E. Munk, "A Computerized Infrared Spectral Interpreter as a Tool in Structural Elucidation of Natural Products," *J. Org. Chem.,* **42**, 1761–1767 (1977).

Woodruff, H. B., and G. M. Smith, "Computer Program for the Analysis of Infrared Spectra," *Anal. Chem.,* **52**, 2321–2327 (1980).

Woodruff, H. B., and G. M. Smith, "Generating Rules for PAIRS: A Computerized Infrared Spectral Interpreter," *Anal. Chim. Acta,* **133**, 545–553 (1981).

Woodruff, H. B., and C. M. Smith, "PAIRS: Program for the Analysis of Infrared Spectra," QCPE Program No. 426.

Ying, L. S., P. Levine, S. A. Tomellini, and S. R. Lowry, "Self-Training, Self-Optimizing Expert System for Interpretation of the Infrared Spectra of Environmental Mixtures," *Anal. Chem.* **59**, 2197–2203 (1987).

17

GRAPHICAL DISPLAY OF DATA

Over the past 30 years there has been an explosion in the ability of science to generate experimental data because of increases in instrumental sophistication, computerization, and related events. In addition, the widespread availability of computers has made it possible to work on scientific problems of a magnitude not approachable previously, that is, experiments where the volume of data necessary to characterize the phenomenon is simply too large to permit its gathering and handling by any means other than by computer. The amount of information that pours out of modern laboratory instrumentation or out of large computational simulations is simply enormous. This fact places a great burden on scientists in two ways: (1) presentation and clarification of the results to themselves so that they can attempt to understand the meaning of the results and (2) presentation of the results (often very much pruned and selected) to others so that information can be transmitted. One of the time-honored traditions of scientists is that of producing pictorial forms called *graphs* for presentation of information to ourselves or to a wider audience. Plots of variables in a graphical representation are nearly always more informative than large tables of data. Humans can perceive patterns in plotted data with little effort, whereas making sense of large sets of tabular material is nearly impossible. In many situations, a data set can be successfully analyzed by using only graphical methods. In other cases, graphical representations can enhance other forms of analysis.

The types of graphical representations that can be used to present data are completely dependent on the quantity and type of data to be presented. The simplest plots show how one dependent variable's value changes as a function of the value of the independent variable. However useful such simple plots

18

GRAPHICAL DISPLAY
OF MOLECULES

An important capability of computers is presenting complex information in pictorial form. In the context of chemistry in general and of molecular mechanics and the study of molecular conformations in particular, this translates into the ability to display molecular structures. Presentation of molecules for visual analysis allows the chemist to view the model and judge its quality or seek insights based on the structure.

People are extremely accomplished at interpreting pictorial images because our primary information source is visual. Digital computers are extremely accomplished at performing routine computations tirelessly. Thus, graphical display of chemical information is a good example of human–machine cooperation. The machine does the repetitive work of managing the molecular model and displaying it, and the person views the result with higher-level objectives in mind. The importance of molecular displays to modern chemistry is emphasized by the number of computer-generated views that appear in textbooks and journal articles and the existence of publications devoted to graphics, such as the *Journal of Molecular Graphics*.

Viewing of objects on computer graphics display terminals in three dimensions is inherently more complex than the more common two-dimensional plots. The display screens are two-dimensional themselves. The solution to the mismatch between the two-dimensional screen and the three-dimensional molecular structure to be displayed on it is the use of projections. A projection is used to transform the three-dimensional molecule onto a two-dimensional projection plane. This projection plane can then be displayed directly.

A molecular structure consists of the identities of the atoms and the spatial coordinates of each atom. The coordinates either will be available in the

Cartesian system or can be transformed into it. Thus, the x, y, and z coordinates of each atom are specified. To map such a three-dimensional object into two-dimensional display, one alternative is to ignore the z coordinates of the atoms, that is, let them all be zero. This is equivalent to making what is called a *parallel projection,* which is the view you get of a solid object when your eye is distant from the object, that is, the perspective is not apparent. In this instance, your line of sight would be parallel to the z axis. This is in distinction to perspective projections used in graphical display of objects in computer-aided design and engineering applications. For a complete discussion of these points, see a computer graphics text (e.g., Foley 1982).

The simplest displays of molecules are skeletal, stick, or wire-frame models. The molecule's bonds are represented by lines, and the atoms are understood to be present at the bond junctions. This is the same kind of graphical representation of molecules that is used in sketching molecular structures on blackboards or paper (hard copy). This type of display can easily be programmed, but the picture generated can be confusing for all but quite simple structures. Figures 18.1*a* and 18.1*b* show a substituted norbornane molecule displayed in stick representations without and with hydrogen atoms to show how congested

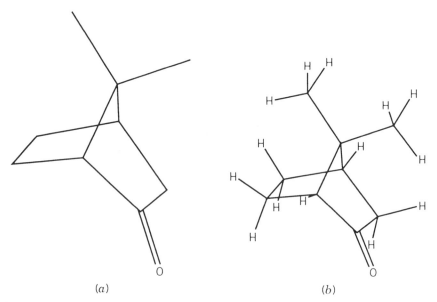

(*a*)   (*b*)

Figure 18.1   Substituted norbornane molecule drawn in a number of representations for comparison: (*a*) stick form with hydrogen atoms suppressed; (*b*) stick form with hydrogen atoms included; (*c*) space-filling atoms with bonds also shown in an ORTEP-like representation; (*d*) space-filling atoms with intersections shown to form a CPK-like representation with hydrogen atoms suppressed; (*d*) space-filling atoms with intersections shown to form a CPK-like representation with hydrogen atoms included.

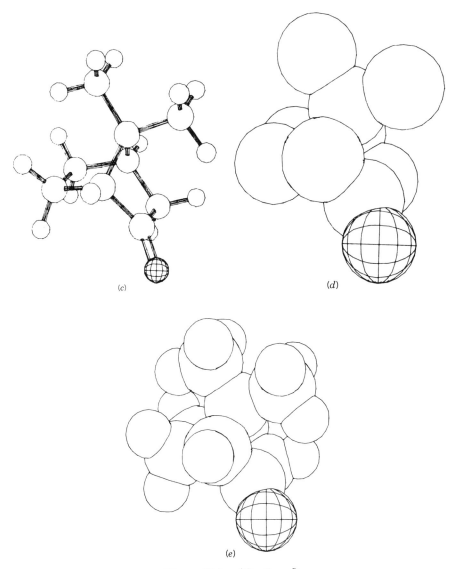

(c)

(d)

(e)

Figure 18.1. (*Continued*)

and confusing a stick display can become, even for a relatively simple molecule. With stick models it is quite possible, even easy, to lose track of which atoms are in the front of the molecular model and which are in the back.

Stick models can be made to appear more three-dimensional, and therefore realistic, in several ways. Slow rotation of the stick model is feasible with some types of computer terminals, and this improves the clarity of the image.

Such rotation gives the viewer a sense of the three-dimensionality of the image. Drawing of stereo pairs is also possible. Here, two copies of the same image are drawn side by side on the display terminal, differing from each other only in the degree of rotation about the vertical axis. Then the user can combine the two images either by using stereo glasses or by defocusing the

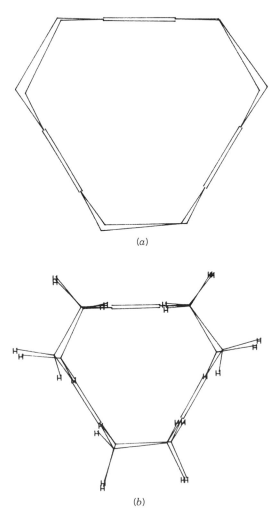

(a)

(b)

Figure 18.2   Tetracyclooctadecatriene drawn in a number of representations for comparison: (a) stick form with hydrogen atoms suppressed; (b) stick for with hydrogen atoms included; (c) space-filling atoms in CPK-like representation with hydrogen atoms suppressed; (d) space-filling atoms in CPK-like representation with hydrogen atoms included; (e) space-filling atoms in CPK-like representation with hydrogen atoms included and cross-hatching to suggest three-dimensionality.

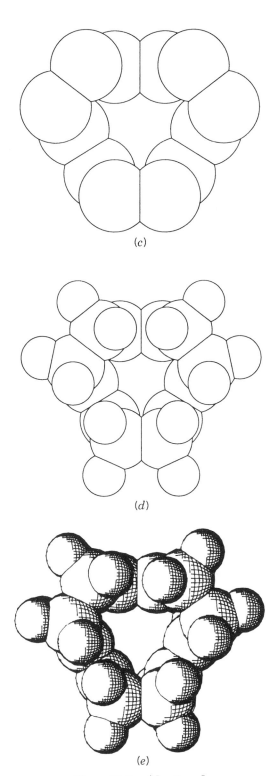

(c)

(d)

(e)

Figure 18.2. (*Continued*)

eyes. Although these displays do demonstrate molecular architecture, they fail to convey information about the space-filling characteristics of the atoms. They provide little indication of how closely the nonbonded atoms approach each other.

A substantial increase in the information content of a molecule is obtained when the atoms are represented by balls and the bonds by sticks. This type of molecular drawing is extremely common in x-ray crystallography, and most papers reporting research on crystallography present graphical displays. Perhaps the best known display of this type is known as ORTEP (Oak Ridge thermal ellipsoid plot) (Johnson 1976). This display uses a graphics display

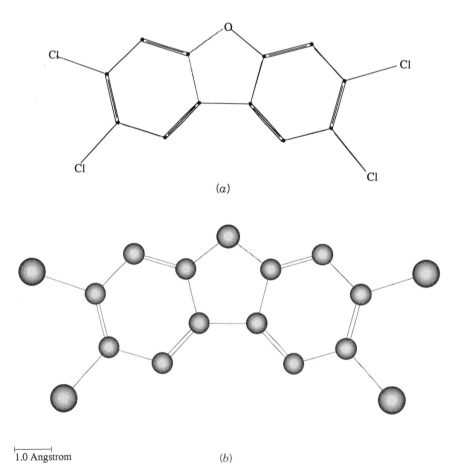

(a)

(b)

1.0 Angstrom

Figure 18.3 2,3,6,7-Tetrachlorodibenzodioxin drawn in four representations with hydrogens suppressed for comparison: (a) stick form; (b) ORTEP-like plot; (c) space-filling atoms with atomic radii adjusted so atoms just intersect; (d) space-filling atoms with atomic radii set at van der Waals radii.

terminal or hard-copy device that will draw lines in one color. Figure 18.1c illustrates this type of representation. There is no doubt about which atoms are in front and which are in back because there are cues indicating how the bonds are drawn and also how the front atoms obscure the rear ones.

An alternative approach for the display of molecules involves representing each atom by a sphere. Lines are drawn where the atoms intersect, and hidden lines are suppressed. This type of display shows a molecule as a three-

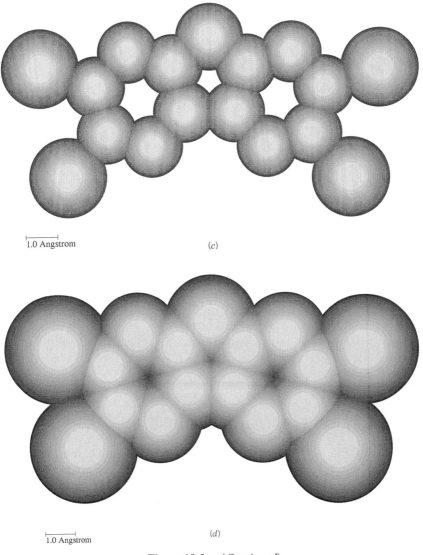

1.0 Angstrom   (c)

1.0 Angstrom   (d)

Figure 18.3.   (*Continued*)

dimensional space-filling object. This type of display also uses lines only for the representation. Figures 18 1d and 18.1e show the substituted norbornane molecule in this representation. This display makes clear how large a volume the molecule occupies and how close together the nonbonded atoms are. The CPK-like figures show how crowded the two acyclic methyl groups are in the norbornane molecule.

The visual display of molecular structures assumes even more importance for larger, more complex molecules. It becomes a virtual necessity as the molecular structures become more and more complex and more inherently three-dimensional. There is a class of compounds composed of a number of six-member rings joined together by double bonds to form large overall rings. The second member of this class, with three six-member rings, was reported in a paper by McMurry et al. (1984). It is a tetracyclooctadecatriene compound, and it is quite difficult to sketch on a blackboard. However, the use of computer graphics displays allows it to be displayed so that the overall structure is easily seen. Figures 18.2a–18.2e show this compound in several views as a stick figure and as a space-filling figure. The final figure, Figure 18.2e, shows how simple cross-hatching of the space-filling drawing can enhance the perception of three-dimensionality.

The availability of laser printers coupled with personal computers has led to the widespread use of molecular displays with atoms represented as spheres. A hierarchy of displays of the same molecule, 2,3,7,8-tetrachlorodibenzodioxin, is shown in Figures 18.3a–18.3d.

In addition to providing the user with visual images of molecules, computer-generated displays provide some other advantages. Any function that depends on three-dimensional model characteristics can be evaluated. For example, through-space distances between atoms that are not bonded to each other are readily available. Such distances could be computed as a function of conformational change. Second, two or more images can be superimposed on the screen. One of them could be rotated or otherwise manipulated while being compared to the other. A third example would be the increase in utility found if a molecular display system were linked to a database of molecular structures, thereby allowing the user to browse through large sets of compounds seeking features of interest.

Display of Biomolecules and Macromolecules

Additional information can be included in a display when color is used. The advent of color display terminals and their incorporation into drug design and biomolecule research has led to an increase in the number of research articles that are accompanied by color space-filling displays (e.g., Knowlton and Cherry 1977, Max 1979, Lesk and Hardman 1982, Feldmann 1983, Langridge et al. 1981, Bash et al. 1983). Max has implemented a routine that simulates the surface of molecules to give a visual impression of the overall surface of a structure. Langridge's group have developed a system to represent molecular surfaces with dots of color, and they have used this display to study interactions between biomolecules. An example of the publication of a color display in a research article whose focus is on structure–activity relations is given in Hansch et al. (1982).

REFERENCES

Bash, P. A., N. Pattabiraman, C. Huang, T. E. Ferrin, and R. Langridge, "Van der Waals Surfaces in Molecular Modelling: Implementation with Real-Time Computer Graphics," *Science,* **222,** 1325–1327 (1983).

Feldmann, R. J., "The Design of Computer Systems for Molecular Modeling," *Ann. Rev. Biophys. Bioeng.,* **5,** 447–510 (1976).

Feldmann, R. J., "Directions in Macromolecular Structure Representation and Display," in *Computer Applications in Chemistry,* S. R. Heller and R. Potenzone (eds.), Elsevier Amsterdam, 1983.

Foley, J. D., *Fundamentals of Interactive Computer Graphics,* Addison-Wesley, Reading, MA, 1982.

Hansch, C., R. L. Li, J. M. Blaney, and R. Langridge. "Comparisons of the Inhibition of *Escherichia coli* and *Lactobacillus casei* Dihydrofolate Reductase by 2,4-Diamino-5-(Substituted-benzyl) Pyrimidines: Quantitative and Structure Activity Relationships, X-Ray Crystallography, and Computer Graphics in Structure-Activity Analysis," *J. Med. Chem.,* **25,** 777–784 (1982).

Hassall, C. H., "Computer Graphics as an Aid to Drug Design," *Chem. Br.,* **21,** 39–46 (1985).

Hopfinger, A. J., "Computational Chemistry, Molecular Graphics and Drug Design," *Pharm. Internatl.* **5,** 224–228 (1984).

Humblet, C., and G. R. Marshall, "Three-Dimensional Computer Modeling as an Aid to Drug Design," *Drug Devel. Res.,* **1,** 409–434 (1981).

Johnson, C. K., "ORTEP-II: A Fortran Thermal-Ellipsoid Plot Program for Crystal Structure Illustrations," ORNL-5138, Oak Ridge National Laboratory, TN, March 1976.

Knowlton, K., and L. Cherry, "ATOMS: A Three-D Opaque Molecule System for Color Pictures of Space-Filling or Ball-and-Stick Molecules," *Comput. Chem.,* **1,** 161 (1977).

Kollman, P., "Theory of Complex Molecular Interactions: Computer Graphics, Distance Geometry, Molecular Mechanics, and Quantum Mechanics," *Accts. Chem. Res.,* **18,** 105–111 (1985).

Langridge, R., T. E. Ferrin, I. D. Kuntz, and M. L. Connolly, "Real-Time Color Graphics in Studies of Molecular Interactions," *Science,* **211,** 661–666 (1981).

Lesk, A. M., and K. D. Hardman, "Computer-Generated Schematic Diagrams of Protein Structures," *Science,* **216,** 539–540 (1982).

McMurry, J. E., G. J. Haley, J. R. Malt, J. C. Clardy, G. Van Duyne, R. Gleiter, and W. Schafer, "Tetracyclo $(8.2.2.2^{2/5}.2^{6.9})$octadeca-1,5,9-triene," *J. Am. Chem. Soc.,* **106,** 5018–5019 (1984).

Max, N. L., "ATOMLLL: A Three-D Opaque Molecule System (Lawrence Livermore

Laboratory Version)," UCRL-52645, Lawrence Livermore National Laboratory, Jan. 1979.

Ramdas, S., and J. M. Thomas, "Computer Graphics in Solid-State and Surface Chemistry," *Chem. Br., 21*, 49–52 (1985).

Smith, G. M., and P. Gund, "Computer-Generated Space-Filling Molecular Models," *J. Chem. Inf. Comput. Sci., 18*, 207 (1978).

Vinter, J. G., "Molecular Graphics for the Medicinal Chemist," *Chem. Br., 21*, 32–38 (1985).

Warme, P. K., "Space-Filling Molecular Models Constructed by Computer," *Comput. Biomed. Res., 10*, 75–82 (1977).

INDEX